CAMBRIDGE LIBRARY COLLECTION

Books of enduring scholarly value

Technology

The focus of this series is engineering, broadly construed. It covers technological innovation from a range of periods and cultures, but centres on the technological achievements of the industrial era in the West, particularly in the nineteenth century, as understood by their contemporaries. Infrastructure is one major focus, covering the building of railways and canals, bridges and tunnels, land drainage, the laying of submarine cables, and the construction of docks and lighthouses. Other key topics include developments in industrial and manufacturing fields such as mining technology, the production of iron and steel, the use of steam power, and chemical processes such as photography and textile dyes.

The Life of John Metcalf

Blinded by smallpox at the age of six, John Metcalf (1717–1810) led a life that might have featured in an eighteenth-century novel. Popularly known as 'Blind Jack of Knaresborough', Metcalf had many and varied careers, including musician, horse trader, fish supplier, textile merchant and stage-wagon operator. Developing a method for building roads on marshy ground, using heather and gorse as a foundation, he eventually became one of the eighteenth century's great road builders, laying over 120 miles of high-quality roads in Yorkshire, Lancashire, Derbyshire and Cheshire. Published in 1795 and based on conversations with Metcalf, this book recounts his life in a series of anecdotes. Metcalf starts with his boyhood escapades and his becoming an accomplished swimmer, climber and gambler. Among the later episodes recounted are his service in raising troops to fight Jacobite rebels, during which he was present at the battles of Falkirk Muir and Culloden.

Cambridge University Press has long been a pioneer in the reissuing of out-of-print titles from its own backlist, producing digital reprints of books that are still sought after by scholars and students but could not be reprinted economically using traditional technology. The Cambridge Library Collection extends this activity to a wider range of books which are still of importance to researchers and professionals, either for the source material they contain, or as landmarks in the history of their academic discipline.

Drawing from the world-renowned collections in the Cambridge University Library and other partner libraries, and guided by the advice of experts in each subject area, Cambridge University Press is using state-of-the-art scanning machines in its own Printing House to capture the content of each book selected for inclusion. The files are processed to give a consistently clear, crisp image, and the books finished to the high quality standard for which the Press is recognised around the world. The latest print-on-demand technology ensures that the books will remain available indefinitely, and that orders for single or multiple copies can quickly be supplied.

The Cambridge Library Collection brings back to life books of enduring scholarly value (including out-of-copyright works originally issued by other publishers) across a wide range of disciplines in the humanities and social sciences and in science and technology.

The Life of John Metcalf

Commonly Called Blind Jack of Knaresborough

JOHN METCALF

CAMBRIDGE
UNIVERSITY PRESS

CAMBRIDGE
UNIVERSITY PRESS

University Printing House, Cambridge, CB2 8BS, United Kingdom

Cambridge University Press is part of the University of Cambridge.
It furthers the University's mission by disseminating knowledge in the pursuit of
education, learning and research at the highest international levels of excellence.

www.cambridge.org
Information on this title: www.cambridge.org/9781108079136

This edition first published 1795
This digitally printed version 2017

ISBN 978-1-108-07913-6 Paperback

JOHN METCALF AGED 79.

Drawn by J. R. Smith.

THE

L I F E

OF

JOHN METCALF,

COMMONLY CALLED

Blind Jack of Knaresborough.

WITH

Many Entertaining ANECDOTES of his EXPLOITS in
Hunting, Card-Playing, &c.

Some PARTICULARS relative to the

Expedition againſt the REBELS in 1745,

IN WHICH HE BORE A PERSONAL SHARE.;

AND ALSO

A Succinct ACCOUNT of his various CONTRACTS for

Making ROADS, Erecting BRIDGES,

AND OTHER UNDERTAKINGS,

IN

*Yorkſhire, Lancaſhire, Derbyſhire,
and Cheſhire ;*

Which, for a Series of Years, have brought him into
PUBLIC NOTICE, as a moſt

EXTRAORDINARY CHARACTER.

EMBELLISHED WITH
A STRIKING HALF-LENGTH PORTRAIT.

YORK:
PRINTED BY E. AND R. PECK, LOW-OUSEGATE.
1795.

[*Entered at Stationers' Hall.*]

ADVERTISEMENT.

TO a generous public little apology will be neceſſary for offering to their patronage the Story of an Individual, who, under circumſtances the moſt depreſſing in their nature, has been, for a conſiderable part of a long life, their aſſiduous and uſeful ſervant.

The Blind, in all ages and countries, have engaged, in a peculiar degree, the ſympathy of mankind ;—and, where original poverty has been annexed to their misfortune, it has been eſteemed the utmoſt exertion in their favour, to enable them to miniſter to the amuſement of ſociety, as the only means for keeping them independent of it : To this general rule, however, a ſurpriſing exception is here ſhewn ; and it is fortunate for the credibility of this little piece, that it is given to the world during the life-time not only of its HERO,

but

but of many others who were witnes-
fes of the various extraordinary *facts*
it contains.

It is fit, however, to notice the
difadvantages under which it now
makes it appearance ;—and which,
from circumſtances, were unavoid-
able: The perſon whoſe taſk it was
to render it, in ſome degree, fit for
the preſs, had much difficulty to en-
counter in arranging the dates, ſcarce
any attention having been paid to
chronological order; and the various
anecdotes having been ſet down, as
the recollection of them aroſe in the
mind of the narrator, by an amanuen-
ſis wholly unqualified for the purpoſe,
and given in a language intelligible to
thoſe only who are well acquainted
with the Yorkſhire dialect.—To thoſe
inaccuracies was added, a literal *dif-
reſpect of perſons;* the firſt and third
being indiſcriminately uſed throughout. To avoid conſtantly-recurring
egotiſms,

egotifms, the preference is here given to the third perfon; though it is feared even that will be found too often in the proper name, where it might have been, in many inftances, fupplied by the pronoun.——But a long abfence having neceffarily fufpended the attention of the Editor, and the defire for publication before the clofe of the Harrogate feafon being urgent, he is not allowed time to correct his own errors. For the fame reafon, the part containing an account of the fhare which Metcalf bore under the late Colonel Thornton, in his expedition againft the rebels; his various undertakings as a road-maker, &c. have received little other correction than what could be given by interlineation. Throughout, however, not the leaft violence is done to the facts; to infure the purity of which, the M.S. has been read over to Metcalf himfelf, and corrected by his defire, wherever any little accidental error has appeared.

Though

Though it was abfolutely neceffary to bring the ftyle into fomething like grammatical order, and to purge it of barbarifms, yet pains have been taken to preferve its fimplicity; and in fome inftances, where a few fentences of dialogue are introduced, the original words remain. Imperfect as it is, a hope is neverthelefs entertained that it will prove amufing; and happy fhall the Author of its Apology be, if the profits arifing from the fale fhall prove of fufficient value to fmooth the decline of a life, which, though marked by eccentricity, has not been fpent in vain.

THE

THE
LIFE

OF

JOHN METCALF.

JOHN METCALF was born at Knarefborough, on the 15th of Auguſt, 1717.
When four years old, he was put to ſchool
by his parents, who were working people,
and continued at ſchool two years: He was
then ſeized with the ſmall-pox, which rendered him totally blind, though all poſſible
means were uſed to preſerve his ſight.

About ſix months after recovering from
the ſmall-pox, he was able to go from his
father's houſe to the end of the ſtreet, and
return, without a guide; which gave him

A much

much fpirit and fatisfaction.—In the fpace of three years he was able to find his way to any part of the town of Knarefborough; and had begun to affociate with boys of his own age, going with them to feek birds' nefts, and for his fhare of the eggs and young birds he was to climb the trees, whilft his comrades waited at the bottom, to direct him to the nefts, and to receive what he fhould throw down; and from this he was foon able to climb any tree he was able to grafp. He would now ramble into the lanes and fields alone, to the diftance of two or three miles, and return. His father keeping horfes, he learned to ride, and in time became an able horfeman, a gallop being his favourite pace. His parents having other children, at the age of thirteen had John taught mufic, at which he proved very expert; though he had conceived more tafte for the cry of a hound or a harrier, than the found of any inftrument.

A gentleman at Knarefborough, of the name of Woodburn, was mafter of a pack of hounds:—This gentleman encouraged

Metcalf

Metcalf very much, by taking him to hunt with him, and was always very defirous of his company. Metcalf kept a couple of very good hounds of his own.

Mr. Woodburn's hounds being feldom kennelled, Metcalf ufed to take feveral of them out fecretly along with his own, about ten or eleven o'clock at night, (the hares being then feeding); but one of the young hounds happening to worry a couple of lambs, it caufed him to difcontinue that practice.

When about fourteen years old, his activity of limbs, and the good fuccefs with which his exploits were ufually attended, confoled him fo greatly for the deprivation of fight, that he was lead to imagine it was in his power to undertake any thing, without danger: the following adventure, however, caufed him to alter his opinion of its value.

There happened to be a plumb-tree a little way from Knarefborough, where there had been a houfe formerly.—One Sunday, Metcalf and his companions (who were fkilled in

matters

matters of this fort) would go there, to get
fome of the fruit; in thefe cafes, Metcalf
was always appointed to afcend, for the purpofe of fhaking the trees. He was accordingly fent up to his poft; but in the height
of the bufinefs, his companions gathering
below were fuddenly alarmed by the appearance of the owner of the tree, and prepared
to quit the ground with all expedition :—
Metcalf thus left to himfelf, foon underftood
how matters were going, though the wind
was high, which prevented him from hearing
diftinctly; and being inclined to follow his
comrades, in making his retreat he fell head-
long into a gravel-pit belonging to Sir Harry
Slingfby, and cut a large gafh in his face,
without, however, receiving any other injury
than a ftun which for fome time hindered his
breathing, and kept him motionlefs on the
ground.——His father being rather fevere,
Metcalf was afraid to go home, left his wound
fhould lead to a difcovery of the prank he
had been engaged in.

Soon

Soon after this, (though not eafily dif-
mayed) he and fome other boys were completely alarmed : — The church - porch at
Knarefborough being the ufual place of their
meeting, they one night between eleven and
twelve o'clock affembled there ; Metcalf
being generally the chief projeétor of their
plans : They determined to rob an orchard ;
which having done, they returned to the
church-porch to divide their booty. Before
their return, a circumftance had happened
to which they were ftrangers, but to the
difcovery of which the following little incident led, though not immediately : There
being a large ring to the church-door, which
turned for the purpofe of lifting the latch,
one of the party took hold of it, and, by way
of bravado, gave a loud rap ; calling out,
" *A tankard of ale here !*" A voice from
within anfwered, very loudly, " *You are at
the wrong houfe.*" This fo ftupified the
whole covey, that none of them could move
for fome time. At length, Metcalf faid,
" Did you not hear fomething fpeak in the
church ?"

church?" Upon this, they all took to their
heels, and ran till they got out of the church-
yard, Metcalf running as faft as any of them.
They now held a confultation on the fubject
of their fright, all equally wondering at the
voice, and none able to account fatisfactorily
for it—One fuppofed that it might have been
fome brother wag, who had put his mouth
to the key-hole of the North door; but to
this it was objected, that the reply was too
diftinct and too ready to have come in that
way. At length, however, their fpirits being
a little raifed, they ventured again down the
flagged pavement into the church-yard; but
when they came oppofite to the church, they
perceived a light, fo great as inclined them
to believe that the church was on fire. They
now re-entered the church-porch, and were
nearly determined to call the parfon; when
fomebody within lifting the latch and making
a great noife, they again difperfed, terrified
and fpeechlefs. One of the party, (whofe
name was Clemifhaw) a fon of the fexton,
ran home, and in a defperate fright got into
bed

bed with his mother; all the reft, at the fame time, making the beft of their way.

The caufe of this panic was as follows :— An old lady, wife of Dr. Talbot, (who had for many years enjoyed the living of Spof-forth) dying, and her relations, who lived at a great diftance, being defirous to arrive before her interment, ordered the body to be kept; this being too long the cafe, and the neighbours perceiving a difagreeable fmell, a requeft was fent to the Rev. Mr. Collins, who ordered the fexton to be called up to dig the grave in the church imme-diately: the fexton had lighted a great number of candles: fo much for the fuppofi-tion of the church being on fire; and the grave-digger was the perfon whofe voice had fo terrified the apple-merchants, when they knocked. Such, however, was the impreffion, that pranks of this nature were not repeated.

About the year 1731, Metcalf being then fourteen years of age, a number of men and boys made a practice of fwimming in the

the river Nidd, where there are many deeps convenient for that purpose.—Metcalf resolving to learn that art, joined the party, and became so very expert, that his companions did not chuse to come near him in the water, it being his custom to seize them, send them to the bottom, and swim over them by way of diversion.

About this time, a soldier and another man were drowned in the above deeps: the former, it was supposed, was taken with the cramp; the latter could not swim. Metcalf was sent for to get up the bodies, and at the fourth time of diving succeeded in bringing up that of the soldier, which, when raised to the surface, other swimmers carried on shore; but life had quite left it. The other body could not then be found.

There are very frequent floods in the river Nidd; and it is a remarkable fact, that in the deep places, there are eddies, or some other causes of attraction, which will draw to the bottom any substance, however light, which comes within their sphere of action. Large

pieces

pieces of timber were often feen to be carried down by the floods ; thefe, on coming over the deep places, were ftopped for the fpace of a moment, and then funk. Upon thefe occafions, Metcalf would go down and with the greateft eafe fix ropes to the wood, which was drawn up by fome perfons purpofely ftationed on the banks.

In the year 1732, one John Barker kept an inn at the Weft end of the High Bridge, Knarefborough. This man was a manufacturer of linen cloth, and ufed to bleach his own yarn. At one time, having brought two packs of yarn to the river to wafh, he thought he obferved a number of wool-packs rolling towards him ; but on a nearer view it proved to be a fwelling of the current, occafioned by a fudden and very violent rain in the neighbourhood. He had not time to remove his yarn, fo that it was fwept away, and carried through the arches of the bridge, which ftands on a rock. A little below there is a piece of ftill water, fuppofed to be about twenty-one feet in depth : as foon as the

yarn

yarn got to this, it funk, except a little which
caught the fkirts of the rock in going down.
Metcalf being intimate with Barker, and calling at his houfe a few days after the accident, found him lamenting his lofs. Metcalf
told him that he hoped to recover his yarn
for him, but Barker fmiled at the fuppofed
abfurdity of the propofal: finding, however,
that his friend was refolved on a trial, he
confented. Metcalf then ordered fome long
cart-ropes to be procured, and fixing a hook
at one end, and leaving the other to be held
by fome perfons on the High Bridge, he
defcended, and hooking as much of the yarn
as he could at one time, he gave orders for
drawing up. In this way the whole was recovered, with very little damage.

Some time after this, Metcalf happened to
be at Scriven, at the houfe of one Green, an
innkeeper.—Two perfons then prefent had a
difpute concerning fome fheep which one of
them had put into the penfold. The owner
of the fheep, (one Robert Scaif, a Knarefborough man, and a friend of Metcalf's)
appeared

appeared to be ill treated by the other party, who wifhed to take an unfair advantage. Metcalf perceiving that they were not likely to agree about the damages, bade them good night, faying he was going to Knarefborough, but it being about the dead time of night, he was firmly refolved to do a little friendly bufinefs before he fhould get home. The penfold being walled round, he climbed over, and getting hold of the fheep one by one, he fairly toffed them over the wall : the difficulty of the fervice increafed as the number got lefs, not being fo ready to catch ;—he was not, however, thereby deterred, but fully completed the exploit.

On the return of day, the penfold door being found faft locked, great was the furprife on finding it untenanted, and various the conjectures as to the rogue or rogues who had liberated the fheep ; but Metcalf paft unfufpected, and enjoyed the joke in filence.

He continued to practice on the violin, until he became able to play country dances.

At

At Knaresborough, during the winter season,
there was an assembly every fortnight, at
which he always attended, and went besides
to many other places where there was public
dancing; yet, though much employed in this
way, he still retained his fondness for hunting, and likewise began to keep game cocks.
Whenever he went to a cock-pit, it was his
custom to place himself on the lowest seat,
and always close to some friend who was a
good judge, and who, by certain motions,
enabled him to bet, hedge, &c. If at any
time he heard of a better game cock than his
own, he was sure to get him by some means
or other, though at a hundred miles distance.

A little way from home he had a cock-
walk, and at the next house there chanced
to be another. The owner of the cock at
the latter house supposing that Metcalf's and
his would meet, armed his own cock with a
steel spur; which greatly displeasing Metcalf,
he formed a plan of revenge; and getting
one of his comrades to assist, they procured
a quantity of cabbage-leaves, and fastening
them

them together with ſkewers, they fixed them
againſt the outſide of the windows, that the
family might not perceive the return of daylight; and that they ſhould alſo be priſoners,
theſe aſſociates in roguery walled up the door
with ſtones, and mud-mortar, which they
were aſſiſted in making by the convenience
of a pump which ſtood near. They then
brought water, in tubs, and continued pour-
ing it in great quantities over the new wall,
(which did not reach quite up to the top of
the door-frame) until the houſe was flooded
to a great depth. This done, they made the
beſt of their way home.

In the morning, the people of the houſe
finding their ſituation, and being at no loſs
to ſuppoſe who had been the projector, and
in all probability the leading performer, of
the buſineſs, were no ſooner ſet at liberty,
than they went to a Juſtice, and got a war-
rant for Metcalf; but not being able to prove
the fact, he was, of courſe, diſmiſſed.

His fame now began to ſpread; and when
any

B

any arch trick was done, inquiry was sure to be made where Metcalf had been at the time.

At Bilton, two miles from Knaresborough, there was a rookery, and the boys had made many attempts to take the young ones; but the owner wishing to preserve them, they were prevented. Metcalf determining to make a trial, sent one of his comrades in the day-time as a spy to reconnoitre the position of the nests; and having been informed by him as to this, they set out in the dead of night, and brought away seven dozen and a half, excepting the *heads*, which they left under the trees. The owner of the rooks finding the heads, sent the bellman round, offering a reward of two guineas for discovering the offenders: the secret, however, was kept until long afterwards.

A man at Knaresborough having married a woman who had lived at a farm-house about a mile distant, brought his wife to his own home; and some articles being left in the deserted house, he sent a son he had by a former marriage to bring them away.—
Metcalf

Metcalf being about the fame age as this boy,
chofe to accompany him. When they got
to the place, the boy miffed the key, which
he had loft from his pocket by the way;
and being afraïd to return without his errand,
he confulted Metcalf about what was to be
done. Metcalf was for entering the houfe
at all events; and not being able to procure
a ladder, got a pole, which reached to the
thatch, and having borrowed a rope and a
ftick, he climbed up the pole, and then afcending by the roof to the chimney, he
placed the ftick acrofs, and faftening the
rope to it, attempted to defcend, but finding
the flew too narrow, he threw off his cloaths,
and laying them on the ridge of the houfe,
made a fecond attempt, and fucceeded: he
then opened the door for his companion.
While they were in the houfe, there was a
heavy thunder-fhower, to which Metcalf's
cloathes were expofed, being left upon the
houfe-top: he attempted to get up again, to
fetch them; but the pole by which he had
afcended was now fo wet, that he could not

B 2   climb

climb by it; he was therefore obliged to wait until it dried, when afcending again, he recovered his cloathes. This was confidered by all who heard of it. as a very extraordinary performance by one in his fituation, as well as a great act of friendfhip to his companion.

In the year 1732 Metcalf was invited to Harrogate, to fucceed, as fidler, a poor old man who had played there for 70 years, and who, being borne down by the weight of 100 years, began to play too flow for country dancing. Metcalf was well received by the nobility and gentry, who employed no other fidler, except a boy whom he hired as an affiftant,.when they began to build a longroom at the Queen's Head.

Being once, with his affiftant, at Ripon affembly, they refolved to call the next day at Newby Hall, the feat of 'Squire Blacket; having got acquainted with that worthy family by their frequent vifits to Harrogate. There they ftayed, regaling themfelves, till near night, when they fet out for home.

In

In the way, they had to crofs the river Ure
by a ford, or go round by Boroughbridge or
Ripon, which latter Metcalf was not inclined
to do. They were told that the ford would
be found impaffable, much rain having
fallen. Metcalf, however, was determined
to try ; but on coming to the water-fide, he
found his companion was much in liquor,
and began to doubt of *his* getting over : as
for himfelf, he had no fear, being a good
fwimmer.—So it was agreed that Metcalf
fhould ftrip, and (leaving his cloathes to the
care of his friend) lead his horfe over, and
thereby prove whether or not it was fafe for
his comrade to follow. By this means they
got over, but not before it was dark. He
then began to drefs himfelf, but his waiftcoat
(in which were the three joints of his hautboy) was miffing, as alfo his filver fhoe-
buckles, and feventeen fhillings which fell
from his pocket. This was an unpleafant
accident, but there being no prefent remedy,
they made the beft of their way to Copgrove,
where they refted. Metcalf liftened difi-

gently

gently to the clock, and after some hours, supposing the waters to have abated, (which was the case,) he returned, and found his seventeen shillings on the bank, and a buckle on each side of the water. The waistcoat and hautboy he could never recover, although he carefully drew the deeps with a gardener's iron rake, which he had procured for that purpose at Newby Hall.

Metcalf now bought a horse, and often ran him for small plates. He still continued to be a cocker—often hunted—and sometimes went a coursing; in the evenings he attended to play at the assemblies: finding, from these various pursuits, pretty sufficient employment. Being greatly encouraged by the gentlemen, he began to think himself of that class, excepting that his *rents* failed to come in-half-yearly from his tenants.

About this time there was a long-room built at the Green-Dragon at Harrogate. More music being then wanted, he engaged one Midgeley (one of the Leeds waits) and

his

his fon, as affiftants. Midgeley, fen. being a
good performer, was taken into partnerfhip
gratis; but the fon, and Metcalf's former
affiftant, paid five pounds each premium.
This was done with the approbation of all
the innkeepers, who wifhed to keep Metcalf
at the head of the band.

In the year 1735, Francis Barlow, Efq; of
Middlethorp, near York, who kept a pack of
beagles, was at Harrogate, and liking Metcalf, gave him an invitation to fpend the
winter at Middlethorp, defiring him to bring
his horfe : the invitation was gladly accepted,
and he went out with Mr. Barlow's hounds
twice a week, highly gratified in the enjoy-
ment of his favourite fport. While at Middlethorp, he was invited by Mr. Hebdin, an
eminent mufician, of York, to come to his
houfe, and play, offering him, gratis, any
fervice or inftruction in his power : this kind
offer Metcalf readily accepted, and went to
practice mufic on thofe days when there was
no hunting.

He

He had now completed a vifit of fix
months to the worthy 'Squire of Middle-
thorpe ;—and the hunting feafon being
almoft over, he propofed to his patron to
take a farewell hunt in the forenoon,
intending to proceed to Knarefborough in
the evening.—He accordingly fet out with
the hounds in the morning ; returned with
the 'Squire at noon ; got himfelf and his
horfe well fed and *watered*, and then proceeded to York, to take leave of Mr. Hebdin, previous to his going home. He had
learned to walk and ride very readily through
moft of the ftreets of York ; and as he was
riding paft the George Inn, in Coneyftreet,
Standifh, the landlord, ftopped him, calling
out "What hafte?" Metcalf told him he
was for Knarefborough that night—The
landlord replied, that there was a gentleman
in the houfe who wanted a guide to Harrogate ; adding, "I know you can do that as
well as any one."—"So I can," faid he,
"but you muft not let him know that I am
blind, for perhaps he will be afraid to truft
me."

me."—"*I* fhall manage that," replied Standifh; fo going in, he informed the gentleman that he had procured him a fafe guide.
Pleafed at this, the gentleman requefted that
Metcalf would come in and take a bottle:
this (for an obvious reafon) the landlord objected to on the part of Metcalf, but recommended fome wine at the door; during the
drinking of which, the ftranger got ready,
and they fet off, Metcalf taking the lead.
As they were turning Oufegate corner, a
voice halloed out " *'Squire Barlow's Blind
Huntfman!*" but the gentleman not knowing the meaning of this, they rode brifkly
up Micklegate, through the Bar, turned the
corner to Holgate, and through Poppleton
Field on to Heffay Moor, and fo proceeded
forward, going over Skip-Bridge. (At this
time the turnpike was not made between
York and Harrogate.)

On the North-Weft end of Kirk-Hammer-
ton Moor, the road to Knarefborough joined
the main road which leads to Boroughbridge
by a fudden turn to the left; but Metcalf
cleared

cleared that without any difficulty. When
they came to Allerton-Mauleverer, the ftranger afked whofe large houfe that was on the
right; and was immediately informed by
Metcalf. A little farther on, the road is
croffed by the one from Wetherby to Boroughbridge, and proceeds along by the
high brick wall of Allerton Park. There was
a road leading out of the Park, oppofite to
the gate upon the Knarefborough road,
which Metcalf was afraid of miffing; but
the wind being from the Eaft, and he perceiving a blaft coming through the Park-
gate, he readily turned his horfe to the oppofite gate which leads to Knarefborough.
Reaching out his hand to open it, he felt the
heel, as it is called; and, backing his horfe,
exclaimed " Confound thee! thou always
goes to the gate heel, inftead of the head."
The gentleman obferved to him that his
horfe feemed aukward, and that his own
mare was good at coming up to a gate;
whereupon Metcalf permitted him to perform this office. Darknefs (which had now
come

come on) being no obftruction to him, he
brifkly led the way, refolved that his companion fhould not again fee his face till they
got to Harrogate. As they were going
through Knarefborough, the gentleman propofed a glafs of wine, which Metcalf refufed,
alledging that the horfes were hot, and that
being near their journey's end, it was not
worth while to ftop:—On then they went;
and prefently fome one cried out " *That's
Blind Jack!*"—This affertion, however, was
contradicted by another perfon who could
not clearly identify him; and by this means
the ftranger was kept in the *dark* as effectually as his guide. They then proceeded over
the High Bridge, and up the Foreft Lane,
and then entering the Foreft, they had to
pafs along a narrow caufeway which reached
about one-third of the way to Harrogate.
When they had gone a little way upon the
Foreft, the gentleman faw a light, and afked
what place it was. There were fome rocks
upon the Foreft called Hookfton Craggs,
and near to thefe the ground was low and
fwampy

fwampy in fome places, clofe by which lays
the Leeds road ;—about this part were frequently feen at night, vapours, commonly
called Will-o'-the-wifp. Metcalf took it
for granted that his companion had feen one
of thefe, but for good reafons declined afking
him whereabout the light was ; and to divert
his attention from this objeƐ, afked him,
" Do you not fee two lights ; one to the
right, the other to the left ?" " No," replied
the gentleman ; " I feen but one light, that
there on the right."—" Well then, Sir,"
faid Metcalf, " that is Harrogate." There
were then many tracks, but Metcalf made
choice of that neareft the fence : by the fide
of this path, which is very near Harrogate,
fome larches were planted ; and ftepping-
ftones laid for the convenience of foot-
paffengers : Metcalf got upon this ftony path,
and the gentleman's horfe following, got
one of his hind feet jammed between two
of the ftones : when his horfe was freed,
he afked " Is there no other road ?" " Yes,"
replied Metcalf, " there is another, but
                                                  it

it is a mile about :" knowing at the fame time that there was a dirty cart-way juft at hand, but to which upon fome account he preferred this rugged path.

Arrived at their journey's end, they ftop- ped at the houfe now called the Granby, but found that the oftler was gone to bed.— Metcalf being very well acquainted with the place, led both the horfes into the ftable, and the oftler foon after appearing, he delivered them to his care, and went into the houfe to inquire after his fel- low-traveller, whom he found comfortably feated over a tankard of negus, in which he pledged his guide; but when Metcalf at- tempted to take the tankard, he reached out his hand wide of the mark : however, he foon found it, and drank; and going out again, left to the landlord the opportunity of explaining to his companion what he was not yet fenfible of.—" I think, landlord," faid the gentleman, " my guide muft have drank a great deal of fpirits fince we came here."—" Why, my good Sir, do you think

C   fo ?"—

so?"—" Well, I judge so from the appearance of his *eyes.*"—"*Eyes!* bless you, Sir,"
rejoined the landlord, "do not you know
that he is BLIND?"—" What do you mean
by that?"—"I mean, Sir, that *he cannot
see.*"—"BLIND! Gracious God!!"—" Yes,
Sir; as blind as a stone, by Heaven!"—
" Well, landlord," said the gentleman, "this
is too much: call him in." Metcalf enters.
" My friend, are you really blind?"—" Yes,
Sir; I lost my sight when six years old."—
" Had I known that, I would not have ventured with you for an hundred pounds."—
" And I, Sir," said Metcalf, "would not
" have lost my way for a thousand." This
conversation ended, they sat down, and
drank plentifully. Metcalf had with him a
case containing a new fiddle which he had
just received from London, and the gentleman observing it, desired him to play: the
guide gave him as much satisfaction in this
way, as he had before done in the character
of a conductor; and the services of the
evening were rewarded by a present of two
                                        guineas,

guineas, befides a plentiful entertainment the next day, at the coft of this gentleman, who looked upon the adventure with Metcalf as the moft extraordinary incident he had ever met with.

1736. The Harrogate feafon now commencing, Metcalf, of courfe, refumed his occupation; and, being of a jocular and comic turn, was fo well received at all the inns, that he obtained free quarters for himfelf and horfe.

The Green Dragon at that place was then kept by a Mr. Body, who had two nephews with him; and when the hunting feafon drew near its clofe, thefe with fome other young men expreffed a great defire for a day's fport; and knowing that Mr. Woodburn, the mafter of the Knarefborough pack of hounds, had often lent them to Metcalf for the fame purpofe, they doubted not of the fuccefs of *his* application: being, however, unprovided with hunters, they were obliged to defer the day for near a fortnight before they could be accommodated.

C 2

On

On the evening before the appointed day,
Metcalf went, flushed with hope, to Mr.
Woodburn, requesting him to lend the pack
for the next day. This was a favour out of
his power to grant, having engaged to meet
'Squire Trapps, with the hounds, next
morning, upon Scotton Moor, for the purpose of entering some young fox-hounds.——
Chagrined at this, Metcalf debated with
himself whether the disappointment should
fall to Mr. Woodburn's friends, or his own:
determining that it should not be the lot of
the latter, he arose the next morning before
day-break, and crossed the High Bridge,
near which he had the advantage of the joint
echos of the Old Castle and Belmont Wood.
He had brought with him an extraordinary
good hound of his own, and taking him by
the ears, made him give mouth very loudly,
himself giving some halloos at the same time.
This device had so good an effect, that in a
few minutes he had nine couple about him,
as the hounds were kept by various people
about the shambles, &c. and were suffered to
lay

lay unkennelled. Mounting his horfe, away
he rode with the dogs to Harrogate, where
he met his friends, ready mounted, and in
high fpirits. Some of them propofed going
to Bilton Wood; but this was oppofed by
Metcalf, who preferred the Moor: in faƈt,
he was apprehenfive of being followed by
Mr. Woodburn, and wifhed to be further
from Knarefborough upon that account.

Purfuant to his advice, they drew the
Moor, at the diftance of five miles, where
they ftarted a hare, killed her after a fine
chace, and immediately put up another:—
juft at this moment came up Mr. Woodburn,
foaming with anger, fwearing moft terribly,
and threatening to fend Metcalf to the devil,
or at leaft to the houfe of correƈtion; and,
his paffion rifing to the utmoft, rode up with
an intention to horfewhip him, which Metcalf
prevented, by galloping out of his reach.—
Mr. Woodburn then endeavoured to call off
the hounds; but Metcalf, knowing the fleetnefs of his own horfe, ventured within fpeaking, though not within *whipping*, diftance of

him,

him, and begged that he would permit the
dogs to finish the chace, alledging that it
would spoil them to take them off; and that
he was sure they would (as they actually
did) kill in a very short time. Metcalf soon
found that Mr. Woodburn's anger had begun
to abate; and going nearer to him, pleaded
in excuse. a misunderstanding of his plan,
which he said he thought had been fixed for
the day after. The apology succeeded with
this good natured gentleman, who, giving
the hare to Metcalf, desired he would accompany him to Scotton Moor, whither, though
late, he would go, rather than wholly difappoint Mr. Trapps. The reader, by this
time, knows enough of Metcalf to believe
he was not averse to this proposal; so leaving
the hares with his comrades, and engaging
to be with them in the evening, he joined
his old associate. The day being advanced;
Metcalf objected to the circuitous way of
Harrogate Bridge, proposing to cross the
river Nidd at Holm Bottom; and Mr. Woodburn not being acquainted with the ford,

he

he again undertook the office of guide, and leading the way, they foon arrived at Scotton Moor, where Mr. Trapps and his company had waited for them two hours. Mr. Woodburn explained the caufe of the delay, and, being now able to participate in the joke, the affair ended very agreeably.

Metcalf ftayed with this company until three in the afternoon, and then fet off for Harrogate, croffing the river. He had not tafted food that day; but when he got to his friends, he found them preparing the brace of hares, with many other good things, for fupper; and after fpending many jovial hours, he played country-dances till daylight.

When the Harrogate feafon was over, it was Metcalf's conftant cuftom to vifit at the inns, always fpending the evening at one or other of them. At the Royal Oak (now the Granby) in particular, fcenes of mirth were often going forward; and at thefe he greatly attracted the notice of one of the landlord's daughters.

In

In the summer he used often to run his horse for the petty plates or prizes given at the feasts in the neighbourhood; and on all these occasions, when in her power, she was sure to attend, with her female friends. By frequent intercourse, the lady and Metcalf became very intimate; and this intimacy produced mutual regard and confidence. Her mother being a high-spirited woman, had brought up her daughters, as she hoped at least, with notions ill suited to the condition of Metcalf; so that in order to disguise the state of their hearts from her parents, the lovers agreed on a set of names and phrases, intelligible to each other, though not so to them. He used to call himself Mary, or Tibby, (at once changing the sex, and speaking as if of a third person); and she, Harry, or Dickey, or some such name. Whenever he sought to intimate to her his intention of visiting her, he would say, "You must tell Richard that Mary will be here on such a day." Her mother would perhaps

ask,

afk, " Who is that ?" To which fhe would
reply, that it was a young woman who was
to meet her brother there.—But if the day
appointed by Metcalf was not convenient,
fhe would fay, that " Richard had called,
and had left word that Mary fhould call
again at fuch a time ;" meaning the time fhe
wifhed Metcalf to come.—And as fhe commonly faftened the doors, when fhe expected
him fhe always left a door or a window
open.

One night, in particular, Metcalf having,
in confequence of an appointment, arrived
there about midnight, and got in by a window
that had been defignedly left open ; in his
way to the *young* woman's room, he met the
old one in the middle of the ftair-cafe ! Both
parties were much furprifed ; and the miftrefs
afking angrily " Who's there ?" " What do
you want ?" he knowing that fhe always
went to bed early, replied " I came in late
laft night, fat down in a chair by the fire-
fide, and fell faft afleep." She then called
loudly

loudly to her daughter, " Why did you not
ſhew Jack to bed ?" " I was not to ſit up all
night for him ;" replied the laſs. He then
purſued his way up ſtairs, and the girl conducted him to a bed-room.

In ſummer he would often play at bowls,
making the following conditions with his
antagoniſt, viz. to receive the odds of a bowl
extra for the deficiency of an eye.—By theſe
terms he had three for the other's one. He
took care to place a friend and confidant at
the jack, and another about mid-way ; and
thoſe, keeping up a conſtant diſcourſe with
him, enabled him, by their voices, to judge
of the diſtance. The degree of bias he
could always aſcertain by feeling ; and, odd
as it may ſeem, was very frequently the
winner.

Cards, too, began to engage his attention ;
all of which he could ſoon diſtinguiſh, un-
affiſted ; and many were the perſons of rank
who, from curioſity, played with him, he
generally winning the majority of the games.

But

But the atchievements already enumerated were far from bounding either his ambition or capacity : He now aspired to the acquaintance of jockies of a higher class than he had hitherto known, and to this end frequented the races at York and many other places ; when he always found the better kind of persons inclined to lend him their skill in making his bets, &c. impressed, as they no doubt were, with sympathy for his situation, and surprize at his odd propensity.

He commonly rode to the race-ground amongst the crowd; and kept in memory both the winning and losing horses.

Being much in the habit of visiting York in the winter time, a whim would often take him to call for his horse at bed-time, and set out for Knaresborough, regardless of the badness of the roads and weather, and of all remonstrance from his friends ; yet the hand of Providence always conducted him in safety. —It was quite common for him to go from Skipton, over the Forest Moor, to Knaresborough, alone ; but if he had company,

and

and it was night, he was, of courſe, the foremoſt.

About the year 1738, Metcalf having increaſed his ſtud, and being aware of the docility of that noble animal, the horſe, ſo tutored his own, that whenever he called them by their reſpective names, they would immediately anſwer him by neighing. This was chiefly accompliſhed by ſome diſcipline at the time of feeding. He could, however, without the help of thoſe reſponſes, ſelect his own horſes out of any number.

Having matched one of his horſes, to run three miles, for a wager of ſome note, and the parties agreeing to ride each his own, they ſet up poſts at certain diſtances, on the Foreſt, including a circle of one mile ; having, of courſe, three rounds to go. Great odds were laid againſt Metcalf, upon the ſuppoſition of his inability to keep the courſe. But what did his ingenuity ſuggeſt in this dilemma : or, rather, what did it anticipate ? He procured four dinner-bells from the different inns, with what others he

could

could borrow; and placing a man, with a bell, at each poft, he was enabled, by the ringing, to turn; and fully availing himfelf of the fuperior fleetnefs of his horfe, came in winner, amidft the plaudits and exultations of the multitude, except only thofe who had betted againft him.

A gentleman of the name of Skelton then came up, and propofed to Metcalf a fmall wager, that he could not gallop a horfe of his fifty yards, and ftop him within two hundred. This horfe was notorious as a run-away, and had baffled the efforts of the beft and ftrongeft riders to hold him. Metcalf agreed to the wager, upon condition that he might choofe his ground; but Skelton objected to there being either hedge or wall in the diftance. Metcalf, every ready at any thing that was likely to produce a joke, agreed; the ftakes were depofited; and knowing that there was a large bog near the Old Spa at Harrogate, he mounted at about the diftance of an hundred and fifty yards from it. Having obferved the wind,

D   and

and placed a perſon who was to ſing a ſong, for the guidance of ſound, he ſet off, at full gallop, for the bog, and ſoon fixed the horſe ſaddle-ſkirt deep in the mire. He then floundered through the dirt as well as he was able, till he gained a firm footing; when he demanded his wager, which was allotted him by the general ſuffrage. It was with the greateſt difficulty, however, that the horſe could be extricated.——That Metcalf was ſo well acquainted with this ſpot, was owing to his having, about three weeks before, relieved a ſtranger who had got faſt in it in the night, and whoſe cries had attracted him.

It was now no unuſual thing with him to buy horſes, with a view to ſell them again. Happening to meet with a man who had left the place of huntſman to a pack of ſub-ſcription hounds kept by Sir John Kaye, 'Squire Hawkeſworth, and others, and who had a horſe to ſell, Metcalf inquired his price, at the ſame time requeſting per-miſſion to ride him a little way. Having

trotted

trotted the horfe a mile or two, he returned, telling the owner that the *eyes* of his nag would foon fail. The man, however, ftood firm to his demand of twenty five guineas for the horfe, alledging that he was beauti-fully moulded, only fix years old, and his action good. Metcalf then took the man into the ftable, and defired him to lay his hand upon the eyes of the horfe, to feel their uncommon heat; afking him, at the fame time, how he could, in confcience, demand fo great a price for a horfe that was going blind. This treaty ended with Met-calf's purchafing the horfe, bridle, and fad-dle, for fourteen pounds.

A few days after, as he was riding on his new purchafe, he ran againft a fign-poft, upon the Common, near the Toy-Shop, and nearly threw it down. Not difcouraged by this, he fet off for Ripon, to play at an affembly; and paffing by a place at Harro-gate called the World's-End, he overtook a man going the Ripon road.—With him Metcalf laid a wager of fix-pennyworth of

D 2   liquor,

liquor, that he would get first to an alehouse
at some small distance. The ground being
rough, Metcalf's horse soon fell, and lay for
a while on the thigh of his master, when,
making an effort to rise, he cut Metcalf's
face with one of his fore shoes. The Rev.
Mr. Richardson coming up at this moment,
and expressing his concern for the accident,
Metcalf told him that nothing had hurt him
but the cowardice of his horse, who had
struck him whilst he was down. His instru-
ment, however, suffered so materially, that
he was obliged to borrow one to perform on
for the night, at Ripon, to which place he
got without further accident. The assembly
over, he set off to return to Harrogate, and
arrived there about three in the morning.

He now thought it was time to dispose of
his fine horse, whose eyes began to discharge
much. After applying the usual remedies
of allum blown into the eyes, roweling in
different parts; &c. he found him in market-
able condition ; and knowing that there
would soon be a great shew of horses without

out Micklegate-Bar, at York, he refolved to take the chance of that mart; and fetting out the night before, put up at the Swan, in Micklegate. The next morning, when the fhew began, Metcalf's nag attracted the notice of one Carter, a very extenfive dealer, who afking the price, was told twenty-two guineas. Carter then inquired if he was found, and received for anfwer, "I have never known him *lame;* but I fhall trot him on this pavement, and if there be any ailment of that kind, it will foon appear, with my weight." The dealer bade him fixteen guineas, and a little after, feventeen; which Metcalf, for well-known reafons, was glad to receive.

Having fold his horfe, he fet off on foot for Harrogate; but before he had got to Holgate (about a mile on his way) he was overtaken by a Knarefborough man, on horfeback, who propofed, for two fhillings-worth of punch, to let him ride in turn, dividing the diftances equally. Metcalf thought the man was unreafonable in his

demand,

demand, but agreed to it at length; and giving his companion one tankard, he, by consent, got the firſt ride, with inſtructions to the following effect, viz. That he ſhould ride on till he got a little beyond Poppleton-Field, where he would *ſee* a gate on his right hand, to which he ſhould faſten the horſe, and leaving him for the owner, proceed. Metcalf not *ſeeing* the gate, as deſcribed, rode on to Knareſborough, which was ſeventeen miles from the place where he had left his fellow-traveller. He then left the horſe at the owner's houſe, ſaying that the maſter having got into a return-chaiſe, had deſired him to ride the horſe home.——The owner was greatly enraged at being left to walk ſo long a way; but, on Metcalf's pleading that he never *ſaw* the gate, he found it his intereſt to join in the laugh.

Being now in the prime of life, and poſſeſſing a peculiar archneſs of diſpoſition, with an unceaſing flow of ſpirits, and a contempt of danger, ſeldom if ever equalled by one in his circumſtances, it will not be wondered at

that

that levities, such as are before recited, should have employed a confiderable portion of his time. The fequel, however, will, in due ccurfe, fhew, that he was capable of embarking in, and bringing to perfection, feveral fchemes, of public as well as private utility; and this promife to the reader, it is hoped, will infure his patience, while he is made the companion of the author in a few more of his frolicfome adventures.

In the year 1738 Metcalf attained the age of twenty-one years, and the height of fix feet one inch and an half, and was remarkably robuft withal.

At that time there lived at Knarefborough one John Bake, a man of a ferocious temper and athletic figure. He was confidered in the neighbourhood as a champion, or rather bully; and thus qualified, was often employed *fpecially*, to ferve writs or warrants, in cafes where defperate refiftance was expected. Metcalf going one evening, with a friend, to a public houfe, they there met this Bake; and a fhort time after, Metcalf's

friend

and Bake fat down to cards. The latter
took fome money off the table, to which he
was not entitled; and the former remonftrating on the injuftice of Bake, received
from him a violent blow. Metcalf interpofing
with words only at firft, was treated in the
fame manner; when inftantly entering into
combat with this ruffian, he beftowed upon
him fuch difcipline as foon extorted a cry
for mercy.

To the fame which Metcalf had acquired
by various means, was now added that of a
boxer, though he was far from being ambi-
tious of celebrity in that way. Some little
time after, Metcalf was called up at midnight
by this very Bake, who, knowing by experi-
ence the prowefs and powers of his late antagonift, had prefumed to make a bet of five
guineas, that Metcalf would beat a fellow
whofe company he had juft left.—But Metcalf gave him to underftand, that, although
he had ftore of thumps for thofe who fhould
treat him with infolence, he was no prize-
fighter; and having no quarrel with the man
in

in queftion, he (Bake) might fight or forfeit as he liked beft.

Being defirous of getting a little fifh, he once, unaffifted, drew a net of eighty yards length, in the deepeft part of the river Wharfe, for three hours together. At one time he held the lines in his mouth, being obliged to fwim.

The following wager he laid, and won: He engaged with a man at the Queen's Head at Harrogate, to go to Knarefborough Crofs, and return, in lefs time than the other would gather one hundred and twenty ftones, laid at regular diftances of a yard each, and, taking one ftone at a time, put them all into a bafket placed at one end of the line.

Meeting with fome company, amongft whom there was one of a boaftful turn, Metcalf propofed to go againft him from Harrogate to Knarefborough Crofs, provided he would take the way which Metcalf fhould choofe. To this the other agreed, believing that he could eafily keep pace with Metcalf till he fhould arrive within fight of the Crofs,

and

and that he could then pufh forward, and beat him. But when they got within half a mile of the town, Metcalf quitted the road which leads over the High Bridge, and, knowing that his antagonift could not fwim, made for a deep part of the river above Bridge, and divefting himfelf of his upper drapery, fwam acrofs; at the fame time calling out jeeringly to his adverfary, "that he hoped for the pleafure of his company up to the Crofs." The other, not liking to commit himfelf to the water, gave up the wager.

About this time, Dr. Chambers, of Ripon, had a well-made horfe, which he ufed to hunt; but finding that latterly he became a great ftumbler, he exchanged him with a dealer, who took him to Harrogate, and meeting with Metcalf, told him he had an excellent hunter to fell at a low price.— Metcalf defired to try how the horfe leaped, and the owner agreeing, he mounted him, and found that he could go over any wall or fence, the height of himfelf when faddled. A bargain was foon ftruck; and this happening

pening at the Queen's Head, several gentlemen who were witnesses of the horse's
performance invited Metcalf to accompany
them, two days after, to Belmont Wood,
where a pack of hounds were to throw off.

These hounds were the joint property of
Francis Trapps, Esq; and his brother, of
Nidd, near Ripley. A pack superior to this
was not to be found in the kingdom; nor
were the owners themselves ever excelled in
their attention to their dogs and hunters.

The wished-for day arriving, Metcalf attended the gentlemen, and the hounds were
not long in finding. The fox took away to
Plumpton Rocks, but finding all secure there
he made for Stockeld Wood, and found
matters in the same state as at Plumpton.—He had then run about six miles.
He came back, and crossed the river Nidd
near the Old Abbey, and went on the East
side of Knaresborough, to a place called
Coney-Garths (where there were earths)
near Scriven. Metcalf's horse carried him
nobly; pulling hard, and requiring propor-
tionate

tionate refiftance. The wind being high,
Metcalf loft his hat, but would not ftop to
recover it ; and coming to Thiftle-Hill,
near Knaresborough, he refolved to crofs the
river at the Abbey-Mill, having often before
gone, *on foot*, over the dam-ftones. When
he got to the dam, he attended to the noife
of the fall, as a guide, and ranging his horfe
in a line with the ftones, dafhed forward for
fome part of the way ; but the ftones being
flippery with a kind of mofs, his horfe ftum-
bled, but recovered this and a fecond blun-
der : the third time, however, floundering
completely, away went horfe and rider into
the dam. Metcalf had prefence of mind to
difengage his feet from the ftirrups, during
the defcent ; but both the horfe and himfelf
were immerfed over head in water. He
then quitted his feat, and made for the oppofite fide, the horfe following him. Having
fecured his nag, he laid himfelf down on his
back, and held up his heels to let the water
run out of his boots ; which done, he quickly
re-mounted, and went up a narrow lane
which

which leads to the road between Knarefborough and Wetherby; then through fome
lanes on the North-Eaft fide of Knarefborough; and croffing the Boroughbridge
road, he got to the Coney-Garths, where he
found that the whipper-in only had arrived
before him.

Here the fox had earthed, as was expected;
and the other horfemen (who had gone over
the Low Bridge, and through the town)
after fome time came up.—They were much
furprifed at finding Metcalf there, and attri-
buted the foaked condition of himfelf and
horfe to profufe fweating; nor were they
undeceived till (giving up the fox) they got
to Scriven, where, upon an explanation of
the affair, they laughed heartily.

In the circle of Metcalf's acquaintance at
Knarefborough were two young men, whofe
fifter lived with them in the capacity of
houfekeeper; and fhe being of a jocular
turn, would often, on Metcalf's calling at
the houfe, propofe fuch whimfical fchemes
to him, as gave him reafon to believe that to

E   laugh

laugh and be merry was the chief bufinefs of her life. However, fhe one evening apprifed him of her intention to pay him a vifit in the night, and defired him to leave his door unlocked. A knowledge of the woman's mirthful propenfity made him at firft confider this as a joke; but, on the other hand, he though it *poffible* that a *real* affignation was intended; and being too gallant to difappoint a *lady*, he told her he would obey her orders. Too fure for the future peace of Metcalf, the lady was punctual to her appointment; coming at the dead time of night to his mother's houfe, unawed at paffing by the *church*, whofe fanction was wanting. It would be impertinent to detain the reader on the fubject of the meeting : fuffice it to fay, that Metcalf too had unfortunately left his fcruples at another houfe. In a few months after, this tender creature accofted him in the ufual way—"I am ruined!—undone!—loft for ever, if you do not make an honeft woman of me!—" &c. &c.

What-

Whatever compunction Metcalf might have felt in a cafe of confiding innocence, pleading for the only compenfation in his power, he did not think his confcience very deeply interefted in the prefent : befides, his heart was ftrongly attached to his firft truly refpectable and worthy miftrefs.—His bufinefs, therefore, was to pacify a troublefome client, which he did in the beft manner he was able. The adventure with this dulcinea had happened previous to the above-mentioned hunt ; but when Metcalf accompanied the gentlemen from the Coney-Garths to the village of Scriven, he there heard, on the authority of the landlord of the inn, that a woman had gone that day to filiate a child to him. He endeavoured to be merry on the occafion, alledging, that it could not be fo, as he had not *feen* the woman for feveral years. This produced a laugh among the company ; but with Metcalf it foon took a more ferious turn. On his return to Harrogate he employed his fellow-fidler to procure a meeting between him and his favourite,

Do-

Dorothy Benſon, which was effected with
ſome difficulty; and he took this occaſion
to inform her of his diſgrace, judging it
better to be before-hand with her, in a matter which could not be long concealed.—
" Ah! John," replied ſhe, " thou haſt got
into a ſad ſcrape : but I intreat thee, do not
think of marrying her." Having quieted
the fears of his favourite on that ſcore, he
deſired his aſſiſtant to go with him to Knareſborough, to *ſound the coaſt;* but before they
had got half way, his companion exclaimed,
" Here is the Town-Officer coming !" Metcalf propoſed walking ſmartly on, without
noticing him ; but when they got near, the
Officer, who was a Quaker, called out,
" Stop, I want to ſpeak with thee." He
then explained his errand, and preſſed Metcalf much to marry the woman ; to which
he replied, that he had no thoughts of marriage, and deſired to know whether for
thirty or forty pounds in money the matter
might be made up. " Yea, friend," ſaid
Jonathan, " perhaps I can ſettle the affair

for

for thee on thofe terms." On this, Metcalf
obferved to him, that he muft go to Harrogate, his money being there. The Quaker
agreeing, they went together to a public-
houfe, where Metcalf called for a tankard of
punch, drank part of it, and feeming very
chearful, faid, " I muft go and colle&t my
money : as it is in various hands, perhaps it
will be an hour or more before I can return ;
fo drink your punch, and call for more."
This pretext fucceeding, he left Jonathan to
regale himfelf at his own fuit ; and choofing
the moft private way to a thick wood, he
there fecreted himfelf all day. After fome
hours waiting, the man of the broad brim
loft all patience, and fet out in queft of his
profane ward ; when meeting a gentleman,
he thus accofted him : " Friend! have thee,
perchance, feen a blind fidler ?" The gentleman replied, " I thought that a perfon of
thy cloth had not wanted a fidler." " I tell
thee I want one at this time," quoth the
Quaker ; who, after fome other fruitlefs
inquiries, went home.

E 3

At

At night, Metcalf ventured to break cover; and judging it unfafe to remain in the neighbourhood of the *hounds*, he gave his affiftant directions to put his little affairs in order—then mounting his horfe, he took the road for Scarborough.

As he was walking one day on the fands, with a friend, he refolved to take a fwim in the fea, his companion agreeing to give him an halloo when he fhould think he had gone far enough outward; but the other, not making a fufficient allowance for the noife of the fea, fuffered him to go out of hearing before he fhouted, and Metcalf continued fwimming until he got out of the fight of his friend, who now fufpected he fhould fee him no more. At length he began to reflect, that, fhould he proceed on to Holland, he had nothing in his *pocket* to make him wel-come;—fo turning, and removing his hair from his ears, he thought he heard the breakers beating againft the pier which defends the Spa; finding, by the noife, that

he

he was at a great diftance, he increafed his efforts, and happening to be right, he landed in fafety, and relieved his friend from a very painful fituation.

Having an aunt at Whitby, near the Allum works, he went there, left his horfe, and got on board an allum fhip bound for London.

In London he met with a North-country man who played on the fmall pipes, and who frequented the houfes of many gentle-men in town. By his intelligence Metcalf found out feveral who were in the habit of vifiting Harrogate ; and amongft others, Colonel Liddell, who refided in King-ftreet, Covent-Garden, and who gave him a gene-ral invitation to his houfe. The Colonel was a Member of Parliament for Berwick-upon-Tweed, and lived at Ravenfworth-Caftle, near Newcaftle - upon - Tyne ; and on his return from London into the North, which generally happened in the month of May, he ftopped three weeks at Harrogate, for a number of years fuccefifively.

When

When the winter was over, Metcalf thought he muſt take a *look* out of London. Accordingly he ſet out through Kenſington, Hammerſmith, Colnbrook, Maidenhead, and Reading, in Berkſhire; and returned by Windſor, and Hampton-Court, to London, in the beginning of May. In his abſence, Colonel Liddell had ſent to his lodgings, to let him know that he was going to Harrogate, and that, if agreeable to him, he might go down either behind his coach or on the top. Metcalf, on his return, waited upon the Colonel, and thanked him, but declined his kind offer, obſerving, that he could, with great eaſe, walk as far in a day as he would chooſe to travel. The next day, at noon, the Colonel, and his ſuite, conſiſting of ſixteen ſervants on horſeback, ſet off, Metcalf ſtarting about an hour before them. They were to go by way of Bugden, and he made his way to Barnet. A little way from Barnet the Bugden and St. Albans roads part, and he had taken the latter: however, he made good the deſtined ſtage for ſleeping, which

was

was Welling, and arrived a little before the Colonel, who was furprized at his performance. Metcalf fet off again the next morning before his friends, and coming to Bigglefwade, found the road was croffed with water, there being no bridge at that time. He made a circuitous caft, but found no other way, except a foot-path which he was dubious of trufting. A perfon coming up, afked, "What road are you for?"—He anfwered, "For Bugden." "You have had fome liquor this morning, I fuppofe," faid the ftranger.—"Yes," replied Metcalf; although he had tafted none that day. The ftranger then bid him follow, and he would bring him into the highway. Soon after they came to fome fluices, with planks laid acrofs, and Metcalf followed by the found of his guide's feet; then to a gate, on the fide of the turnpike, which being locked, he was told to climb over. Metcalf was ftruck with the kind attention of his conductor, and taking twopence from his pocket, faid, "Here, good fellow, take that, and get thee

a

a pint of beer;" but the other declined it,
saying he was welcome. Metcalf, however,
preffing the reward upon him, was afked,
"Can you fee very well?" "Not remarkably
well," he replied. " My friend," faid the
ftranger, "I do not mean to *tythe* you:—I
am Rector of this parifh; and fo God blefs
you, and I wifh you a good journey." Metcalf fet forward with the parfon's benedic-
tion, and ftopped every night with the Colonel: On coming to Wetherby, he arrived at
the inn before him, as ufual, and told the
landlord of his approach, who afked him by
what means he had become acquainted with
that, and was informed by him how he had
preceded the Colonel the whole week, this
being Saturday, and they had left town on
Monday noon. The Colonel arriving, ordered Metcalf into his room, and propofed
halting till Monday; but Metcalf replied,
" With your leave, Sir, I fhall go to Harrogate to night, and meet you there on Monday." In truth, he was anxious to know
the worft refpecting the woman who had
been

been the cause of his journey; and was
much pleased to find matters in a better
train than he expected, she being in a com-
fortable way, and not inclined to be farther,
troublesome. Many friends visited him on
Sunday, and the next day the Colonel arrived. But of all his friends, the dearest
was at the Royal Oak: with her he had an
affectionate meeting, after an absence of
seven months. During this interval a young
man had been paying his addresses to her;
and knowing that Metcalf was acquainted
with the family, he solicited him to use what
interest he had in his behalf: this, when
made known to the lady by the man of her
heart, afforded them both great entertainment.

Metcalf became now in great request as a
performer at Ripon assembly, which was
resorted to by many families of distinction,
such as those of Sir Walter Blacket of
Newby, Sir John Wray, Sir R. Graham,
'Squire Rhodes, 'Squire Aiflaby of Studley,
and many others. When he played alone,

it

it was ufual with him, after the affembly, to
fet off for Harrogate or Knarefborough; but
when he had an affiftant, he remained all
night at Ripon to keep him company, his
partner being afraid to ride in the dark.

Finding himfelf worth fifteen pounds, (a
larger fum than he ever before had to fpare)
he made his favourite Mifs Benfon his treafurer; but as he had not yet begun to fpecu-
late in the purchafe of *land,* and a main of
cocks being made in the neighbourhood, he
became a party, and drawing his cafh from
the hands of his fair banker, he loft two-
thirds of his whole fortune.——The remaining
five pounds he laid on a horfe which was to
run at York a few days after; and though
he had the good fortune to win the laft
wager, his general imprudence in this way
produced a little fhynefs from his fweetheart.

His competitor (not fufpecting the intimacy between Metcalf and the young lady)
pufhed his fuit brifkly; and after a fhort
time, banns were publifhed in the churches
of Knarefborough and Kirby-Overblow.——

Metcalf

Metcalf was much furprifed, having long thought himfelf fecure of her affection. He now began to believe that fhe had laid more ftrefs on his late follies than he had been aware of, and the remembrance of them gave him exquifite pain, for he loved her tenderly, and was reftrained from propofing marriage to her only by the doubts he had of being able to fupport her in the manner fhe had been accuftomed to. On the other hand, his pride made him difdain to fhew that he was hurt, or to take any meafures to prevent the match. The publication of banns being complete, the wedding-day was appointed.—The fuppofed bridegroom had provided an entertainment at his houfe for upwards of two hundred people ; and going with a few friends to Harrogate on the Sunday, propofed the following day for the nuptials, which were to be folemnized at Knarefborough, intending to return to Harrogate to breakfaft, where a bride-cake was ready, with a hamper of wine, which latter was to have been carried to Kirby, for the ufe of the guefts he had invited.

On the Sunday, Metcalf riding pretty
fmartly paft the Royal Oak, towards the
Queen's Head, was loudly accofted in thefe
words—" One wants to fpeak with you."
He turned immediately to the ftables of the
Oak, and, to his joyful furprife, found there
his favourite, who had fent her mother's
maid to call him. " Well, lafs," faid he,
" thou's going to have a merry day to-mor-
row; am I to be the fidler ?"—" Thou
never fhalt fiddle at my wedding," replied
fhe. " What's the matter? What have I
done?" faid Metcalf.—" Matters may not
end," faid fhe, " as fome folks wifh they
fhould." " What!" faid he, " hadft thou
rather have me? Canft thou bear ftarving?"
—" Yes," faid fhe, " with *thee* I can !"
" Give me thy hand, then, lafs,—fkin for
fkin, it's all done !"

The girl who had called him being prefent,
he told her, that as fhe and his horfe were
the only witneffes to what had paffed, he
would kill the firft who fhould divulge·it.—
The immediate concern was to fix on fome
plan,

plan, as Mifs Benfon was apprehenfive of being miffed by her friends.—Jack, ever prompt at an expedient, defired that fhe would that night place a lighted candle in one of the windows of the old houfe, as foon as the coaft was clear, and herfelf ready to fet off, which will doubtlefs appear to the reader a very extraordinary fignal to a blind man; but he had conceived meafures for carrying the projected elopement into effect by the affiftance of a third perfon. This being approved of, fhe went into the houfe, and in a fhort time was followed by Metcalf, who was warmly received by the fuppofed bridegroom and company. The tankard went brifkly round with "Succefs to the intended couple;" in which toaft, it may be readily believed, Metcalf joined them moft cordially.

Having ftayed till it was near dark, he thought it time for putting bufinefs into a proper train. Going then to a public houfe known by the name of the World's End, he inquired for the oftler, whom he knew to be

a

a steady fellow; and after obtaining from
this man a promise either to serve him in an
affair of moment in which he was engaged,
or keep the secret, he related the particulars
of his assignation, and the intended elopement; to forward which, he desired him to
let them have his master's mare, which he
knew would carry double.—This agreed on,
he requested the further service of meeting
him at the Raffle Shop (now the Library) at
ten o'clock: a whistle was to be given by
the first who got there, as a signal. They
met pretty punctually; and Metcalf asked
him if he saw a star, meaning the light before
mentioned: he said, he did not; but in less
than half an hour the *star* was in the place
appointed. They then left the horses at a
little distance from the house, not choosing to
venture into the court-yard, it being paved.
On the door being opened by the lady, he
asked her if she was ready; to which she
replied in the affirmative.—He advised her,
however, to pack up a gown or two, as she
probably might not see her mother again for
some

some time. The ostler having recommended the lady's pillion to Metcalf, in preference to that of his mistress, he asked her for it :—— "O dear !" said she, "it is in the other house; but we must have it." She then went to the window and called up her sister, who let her in. The pillion and cloth were in the room where the supposed bridegroom slept; and on his seeing her enter, she said, "I'll take this and brush it, that it may be ready in the morning." "That's well thought on, my dear," said he. She then came down, and all three went to the horses. Metcalf mounted her behind his friend, then got upon his own horse, and away they went. At that time it was not a matter of so much difficulty to get married as it is at present; and they, with only the trouble of riding twelve miles, and at a small expence, were united.

Metcalf left his bride, on his return, at a friend's house within five miles of Harrogate, but did not dismount, being in haste to return the mare he had borrowed with *French leave.* A few minutes after their return, Mr. Body, the landlord, called for his mare, to go to Knaresborough, and fortunately she was ready for him.

Metcalf now went to the Queen's Head, to perform the usual service of playing during the breakfast half hour. His over-night's excursion made him rather thought-ful, having got a *bird*, but no *cage* for it. While he was musing on this subject, an acquaintance, who made one of the intended bridegroom's company the evening before, came up, and asked him to take a glass with him. Metcalf quickly guessed what his busi-ness was, but adjourned with him to a private room, seemingly unconcerned. "Metcalf," said he "a strange thing has happened since you were with us last night, concerning Dolly Benson, who was to have been mar-ried this morning to Anthony Dickinson.—You are suspected of knowing something about the former; and I shall briefly state to you the consternation which her disappear-ance has occasioned, and the reasons why suspicion falls upon you. This morning, early, the bridegroom went to Knaresbo-rough,

rough, and informed the Rev. Mr. Collins that he and his intended wife were coming that forenoon to be married. In his abfence Mrs. Benfon and her other daughter began to prepare for breakfaft; and obferving that Dolly lay very long in bed, her mother defired that fhe might be called: but her ufual bedfellow declaring that fhe had not flept with her, fhe was ordered to feek her in fome of the other rooms. This was done, but in vain. They then took it for granted that fhe had taken a ride with Mr. Dickinfon; but he returning, could give no account of her. All her friends began now to be very ferioufly alarmed; and, amongft other fearful conjectures, fuppofed that fhe might have fallen into the well, in attempting to draw water for breakfaft; and actually got fome iron creepers, and fearched the well. Her brother then took horfe, and rode to Burton-Leonard, to a young man who had flightly paid his addreffes to her, and, informing him of the diftrefs of the family, begged he would give information, if in his power. The young man immediately afked him if he

had

had feen Blind Jack; he anfwered, that you
were at the Oak laft night, but did not in
the leaft fufpect you.—The other, however,
perfifted in the opinion that you were-moft
likely to know where the girl was, and gave
the following incident as a reafon : Being,
not long fince, at a dance, where Mifs Benfon made one, he obferved her wiping a
profufe perfpiration from your face, with an
handkerchief; and this act was accompanied
by a look fo tender, as left no doubt.in his
mind of her being ftrongly attached to you."

This narrative (a part of which was no
news.to Metcalf) was fcarcely finifhed, when
young Benfon appeared; and Metcalf put an
end to all inquiry, by declaring the truth :
and thinking it his duty to conciliate, if
poffible, thofe whom he had offended, he
employed the fofteft phrafes he was mafter
of on the occafion. He begged pardon,
through their fon, of Mr. and Mrs. Benfon,
whom he did not prefume to call father and
mother, and wifhed them to believe that the
warmth of his paffion for their daughter,
with the defpair of obtaining their confent,
had

had led him to the meafures he had taken ; and that he would make them the beft amends in his power, by the affectionate conduct he fhould obferve to his wife.

The fon, in part pacified, left Metcalf, and reported this declaration to his parents : but they were juft as well pleafed at it, as they would have been at the fight of their building in flames ; and, in the height of paffion, declared they would put him to death, if they met with him.

The poor forlorn Dickinfon then departed, accompanied by one of Mr. Benfon's fons. When they got near his home, they heard two fets of bells, viz. thofe of Folifoot and Kirby Overblow, ringing, in expectation of the arrival of the bride and groom ; but the found was more like that of a knell to Dickinfon, who fell from his horfe through anguifh, but was relieved by the attention of his friend. The company were furprifed at not feeing the bride ; but matters were foon explained, and they were defired to partake of the fare provided for them.

Metcalf

Metcalf not being able, at once, to procure a *Palace* for his *Queen*, took a small house at Knarefborough. It now became matter of wonder that fhe fhould have preferred a blind man to Dickinfon, fhe being as handfome a woman as any in the country. A lady having afked her why fhe had refufed fo many good offers for Blind Jack; fhe anfwered, "Becaufe I could not be happy without him:" And being more particularly queftioned, fhe replied, "His actions are fo fingular, and his fpirit fo manly and enterprifing, that I could not help liking him." Metcalf being interrogated, on his part, how he had contrived to obtain the lady, replied, That many women were like liquor-merchants, who purchafe fpirits above proof, knowing that they can *lower* them at home; and this, he thought, would account why many a rake got a wife, while your plodding fons of phlegm were doomed to celibacy.

He now went to Harrogate, as ufual, with the exception of *one* houfe. Meeting with a butcher there one day, and drinking pretty freely,

freely, a wager was propofed to Metcalf, that he durft not vifit his mother-in-law. He took the wager, mounted his horfe, and riding up to the kitchen-door, called for a pint of wine. There being then only women in the houfe, they all ran up ftairs in a fright. He then rode into the kitchen, through the houfe, and out at the hall-door, no one molefting him. As there were many evidences to this act of *heroifm*, he returned; and demanding the ftakes, received them without oppofition.

The Harrogate feafon being on the decline, he retired to Knarefborough, where he purchafed an old houfe, intending to build on its fcite the next fummer. Affifted by another ftout man, he began to get ftones up from the river; and being much ufed to the water, took great delight in this fort of work. Meeting with fome workmen, he told them the intended dimenfions of his houfe, and they named a price, by the rood, for building it: but Metcalf, calculating from his own head, found that their eftimate would

not

not do; fo letting them the job by lump
agreement, they completed it at about half
the fum which they would have got by the
rood.

He now went to the Oak, to demand his
wife's cloaths, but was refufed: on a fecond
application, however, he fucceeded. His
wife having brought him a boy, and fome
genteel people being the fponfors, they employed their good offices to heal the breach
between the families, and were fo fortunate
as to fucceed. On the birth of a daughter
(the fecona child) Mrs. Benfon herfelf was
godmother, and prefented Metcalf with fifty
guineas.

He continued to play at Harrogate in the
feafon; and fet up a four-wheel chaife, and
a one-horfe chair, for public accommodation,
there having been nothing of the kind there
before.—He kept thofe vehicles two fum-
mers, when the innkeepers beginning to run
chaifes, he gave them up; as he alfo did
racing, and hunting; but ftill wanting employment, he bought horfes, and went to
the

the coaſt for fiſh, which he took to Leeds
and Mancheſter; and ſo indefatigable was
he, that he would frequently walk for two
nights and a day, with little or no reſt.

Going from Knareſborough to Leeds in a
ſnow-ſtorm, and croſſing a brook, the ice
gave way under-one of his horſes, and he
was under the neceſſity of unloading to get
him out; but the horſe as ſoon as free ran back
to Knareſborough, leaving him with two pan-
niers of fiſh, and three other loaded horſes,
which, together with the badneſs of the
night, greatly perplexed him :—After much
difficulty, however, he divided the weight
amongſt the others, and purſuing his journey,
arrived at Leeds by break of day.

Once paſſing through Halifax, he ſtopped
at an inn called the Broad Stone. The landlord's ſon and ſome others who frequented
Harrogate ſeeing Metcalf come in, and having often heard of his exploits, ſignified a
wiſh to play at cards with him: he agreed,
and accordingly they ſent for a pack, which
he deſired to examine a little. The man of
G   the

the houfe being his friend, he could depend
upon his honour in preventing deception.
They began, and Metcalf beat four of them
in turn; playing for liquor only. Not fatif-
fied with this, fome of the company propofed
playing for money ; when engaging at fhilling whift, Metcalf won fifteen fhillings.
The party who loft then propofed to play
double or quit, but Metcalf declined playing
for more than fhilling points ; till at laft
yielding to much importunity, he got engaged for guineas, and, favoured by fortune,
won ten, the whole fum late in the poffeffion
of the lofer, who took up the cards, and
going out, foon returned with eight guineas
more : Metcalf's friend examined the cards,
to fee that they were not marked ; and finding all fair, they went on again, until thofe
eight pieces followed the other ten. They
then drank freely at Metcalf's coft, he being
in good circumftances to treat. About
ten at night he took his leave, faying he muft
be at Knarefborough in the morning, having
fent his horfes before. On his way he crof-
sed

fed the river Wharfe about a mile below Poole: the water being high, his horfe fwam, but he got fafe home; and this ended his purfuits as a fifhmonger, the profits being fmall, and his fatigue very confiderable.

From the period of his difcontinuing the bufinefs of fifhmonger, Metcalf continued in the practice of attending Harrogate, as a player on the violin in the Long-room, until the commencement of the Rebellion in 1745.

The events of that period having been fo numeroufly and fo minutely detailed, that any one the leaft converfant in the hiftory of this country cannot be unacquainted with the origin, progrefs, and termination of the civil commotions which agitated it,—it would appear unneceffary to obtrude the narration of them here, further than may feem needful to introduce the part in which Metcalf bore a perfonal fhare. The circumftance of his commencing foldier, was at that time, and will ftill by the reader, be looked upon as a very extraordinary proceeding of one in his fituation.

The alarm which took place, in confequence of that event, was general; and loyalty to the reigning Sovereign, and Government, with meafures for refiftance to the Rebel Party, fhone no where more confpicuous than in the County of York.

Amongft the many inftances which mark this, none were more ftriking than the fignally-gallant conduct of the late WILLIAM THORNTON, Efq; of Thornville.

The opinion of that gentleman, as delivered at the General County Meeting held at the Caftle of York, was, that the four thoufand men, (for the raifing, cloathing, and maintaining of which *ninety thoufand pounds* had been fubfcribed) fhould be embodied in companies with the regulars, and march with the King's forces to any part where their fervices might be required.—This mode of proceeding, however, not meeting the opinion of the majority of the gentlemen prefent, he determined to raife a company at his own expence.

In

In confequence of that refolution, Mr. Thornton went to Knarefborough about the firft of October, 1745; and Metcalf having for feveral years been in the practice of vifiting that gentleman's manfion, (particularly at the feftive feafon of Chriftmas, where, with his violin and hautboy, he affifted to entertain the family) Mr. Thornton was well acquainted with his extraordinary difpofition, and, imagining that he might be of fervice to him in his prefent undertaking, fent for our blind hero to his inn, treated him liberally with punch, and, informing him "that the French were coming to join the Scotch rebels, the confequence of which would be, that if not vigouroufly oppofed, they would violate all our wives, daughters, and fifters," afked him if he had fpirit to join the company about to be raifed. Metcalf inftantly giving an affirmative anfwer, was afked whether he knew of any fpirited fellows who were likely to make good foldiers; and having fatisfied his patron on this head alfo, he was appointed an affiftant to a ferjeant already procured,

G 3   with

with orders to begin recruiting the next
day. This fervice went on with rapid fuccefs: feveral carpenters, fmiths, and other
artificers were engaged, to all of whom
Metcalf promifed great military advance-
ment, or, in default of that, places of vaft
profit under Government, as foon as the
matter was over, which he called only a
buftle ; thus following the example of other
decoy ducks, by promifing very unlikely
things.

Such was their fuccefs, that in two days
only they enlifted one hundred and forty
men, out of whom the Captain drafted fixty-
four, (the number of privates he wanted)
and fent immediately to Leeds for cloth of a
good quality for their cloathing. The coats
were blue, trimmed and faced with buff;
and buff waiftcoats. The taylors he had
employed refufing to work on a Sunday, he
rebuked their fanatical fcruples in thefe
words: "You rafcals! if your houfes were
on fire, would you not be glad to extinguifh
the flames on a *Sunday?*" which had the
de-

defired effect. Arms being procured from
the Tower, the men were conftantly and
regularly drilled. Such of them as had
relations in the public line, would frequently
bring their companions to drink, for the
benefit of the refpective houfes; and Metcalf
never failed to attend one or other of thofe
parties, his fiddle and hautboy contributing
to make the time pafs agreeably : and the
worthy Captain was liberal in his allowance
of money for fuch feftive purpofes, infomuch
that had he wanted five hundred men, he
could eafily have obtained them. Soon after
he brought them to Thornville, where he
ordered every other day a fat ox to be killed
for their entertainment, and gave them beer
feven years old, exprefling a great pleafure
at its being referved for fo good a purpofe.

He now began to found the company as to
their attachment to the caufe and to himfelf.
" My lads," faid he, " you are going to
form a part of a ring-fence to the fineft
eftate in the world! The King's army is on
its march to the Northward; and I have
the

the pleafing confidence that all of you are
willing to join them."—They replied, as if
one foul had animated them, " We will
follow you to the world's end !"

All matters being adjuſted, the company
was drawn up, and amongſt them BLIND
JACK made no *ſmall* figure, being near ſix
feet two inches high, and, like his companions, dreſſed in blue and buff, with a large
gold-laced hat : So well pleaſed was the
Captain with his appearance, that he ſaid he
would give an hundred guineas for only *one
eye* to ſtick in the head of his *dark* champion.

Jack now played a march of the Captain's
chooſing, and off they moved for Boroughbridge. Capt. Thornton having a diſcretion-
ary route, took his march over the moors, in
expectation of meeting ſome of the ſtraggling
parties of the rebel army ; and quartered at
ſeveral villages in his way, where he was
kindly received, and viſited by the heads of
the genteeleſt families in the neighbourhood,
who generally ſpent the evenings with him.
Metcalf being always at the Captain's quarters,

ters, played on the violin, accompanied by a good voice, " *Britons! ſtrike home,*" and other loyal and popular airs, much to the ſatisfaction of the viſitors, who frequently offered him money, but this he always re-fuſed, knowing that his acceptance of it would diſpleaſe his commander.

Arriving at Newcaſtle, they joined the army under the command of General Wade, by whoſe order they were united with Pulte-ney's regiment, which, having ſuffered much in ſome late actions abroad, was thought the weakeſt. Captain Thornton gave orders for tents for his men, and a marquee for himſelf, for which he paid the upholſterer eighty guineas. He pitched them on New-caſtle Moor, and gave a pair of blankets to each tent. Jack obſerved to his Captain, " Sir, I live next door to you : but it is a cuſtom, on coming to a new houſe, to have it warmed." The Captain knowing his meaning, ſaid, " How much will do ?"— Jack anſwered, " Three ſhillings a tent;" which the Captain generouſly gave, and ſaid,

"As

" As you join Pulteney's regiment, they will
fmell your breath;" fo he gave them ten
guineas, being one to each company. On
the night of their entertainment, the fnow
fell fix inches.

After ftopping here for about a week, the
General received intelligence of the motions
of the rebels, and gave orders to march by
break of day for Hexham, in three columns,
wifhing to intercept them upon the Weft
road, as their route feemed to be for England
that way. The tents were inftantly ftruck;
but the Swifs troops having the van, and not
being willing to move at fo early an hour,
it was half paft ten before they left the
ground, and the fnow by that time was
become extremely deep in feveral places: it
alfo proved a very fevere day for hail and
froft. They were often three or four hours
in marching a mile, the pioneers having to
lower the hills, and fill up feveral ditches, to
make a paffage for the artillery and baggage.

About ten at night they arrived at Oving-
ton, the place marked out for them, with
ftraw

ftraw to reft on ; but the ground was frozen
fo hard, that but few of the tent-pins would
enter it, and in thofe few tents which were
pitched, the men lay one upon another,
greatly fatigued with their march, it having
been fifteen hours from the time of their
ftriking the tents, till their arrival at this
place, although the diftance is only feven
miles.

At eleven o'clock at night Captain Thornton left the camp, and went to Hexham, to
vifit his relation, Sir Edward Blacket, and
with a view of getting provifions and necef-
faries for his men : he was only nine hours
abfent, as, although it was Sunday morning,
the march was to be continued. It having
been cuftomary to burn the ftraw, to warm
the men before they fet off, orders were
here given to preferve it, in cafe it might be
wanted on their return. However, Captain
Thornton and the Lieutenant being abfent,
and the Enfign having died at Newcaftle,
Metcalf took it upon him to fay, " My lads,
get the ftraw together, to burn ; our Captain
will

will pay for more, if we fhould want it :" which being done, he took out his fiddle, notwithftanding the day, and played to the men whilft they danced round the fire; which made the reft of the army obferve them, though they did not follow their example. The Captain and Lieutenant arriving in the midft of the bufinefs, expreffed much pleafure and fatisfaction in feeing the men thus recreate themfelves.

That day they reached Hexham, where they halted. On Monday night, about ten o'clock, the army was put in motion by a falfe alarm. Here General Wade refolved to return; and immediately began the march for York, by way of Pierfebridge, Catterick, and Boroughbridge; and continuing his route Southward, encamped his men on Clifford Moor, where they halted a few days, and then moved to a ground between Ferrybridge and Knottingley. The rebels had now penetrated Southward as far as Derby; but the General having heard that they had received a check from the Duke of Cumberland,

land, fent General Oglethorpe with a thoufand horfe towards Manchefter, either to
harrafs the enemy in their retreat, or to join
the Duke's forces; and returned himfelf with
the remainder, by Wakefield-Outwood, and
Leeds, to Newcaftle.

In the mean time the Duke came up with
the rebels at Clifton, on the borders of
Weftmoreland, of which Lord George Murray, with the rear guard, had taken poffeffion,
whilft another party had fortified themfelves
behind three hedges and a ditch.

The Duke coming upon the open moor
after fun-fet, gave orders for three hundred
dragoons to difmount, and advance to the
brink of the ditch; when the rebels fired
upon them from behind the hedges, which
they returned, and fell a few paces back:
the rebels miftaking this for flight, rufhed
over the ditch, but meeting a warmer reception than they expected, were glad to retreat,
and continued their route to Penrith.

The Duke's army was not able to follow,
owing to the badnefs of the roads, and the

fa-

fatigue of a tedious march; but the next morning he purfued them to Penrith; and from thence to Carlifle, where they left part of their army.

His Royal Highnefs thought it advifable to reduce this place, and accordingly fent for heavy artillery from Whitehaven, which arriving on the 25th of December, the garrifon furrendered on the 30th, and his Royal Highnefs returned to London. General Wade continued his march for the North, difmiffing all the foreigners from his army; and General Hawley on coming from London to take the command, was joined by fome regiments which had been withdrawn from Flanders. They marched to Edinburgh; from thence to Falkirk, and pitched their tents on the North-Eaft fide of the town, on the 16th of January.

The Highland army being at Torwood, about mid-way between Falkirk and Stirling, and diftant from the Englifh camp only about three miles, they could eafily difcover each other's camp-lights. The Englifh army lay

all

all night on their arms, in expectation of
being attacked; but the van and picquet
guards came in on the morning of the 17th,
having obferved no motions in the rebel
camp which fhewed any figns of an attack,
although they were as near them as fafety
would permit. Soon after, the enemy were
obferved to move fome of their colours from
Torwood, towards Stirling, which made the
Englifh fuppofe that they were retreating;
but this motion was a feint to deceive them.
However, upon this appearance, the foldiers
were ordered to pile their arms, and take
fome refrefhment; and although Lord Kilmarnock was in the rebel army, General
Hawley went to breakfaft with Lady Kilmarnock, at Callendar-Houfe. The enemy, in
the mean time, ftole a march down a valley
Northward, unperceived; but juft before the
army difcovered them, they were feen by a
perfon who ran into the camp, exclaiming,
" Gentlemen ! what are you about ? the
Highlanders will be upon you:" on which
fome of the officers faid, " Seize that rafcal,

he.

he is spreading a false alarm."—" Will you,
then, believe your own eyes?" replied the
man; when instantly the truth of his assertion
became apparent, by their advancing to the
highest ground upon Falkirk moor, the wind
blowing strongly in the faces of the English,
with a severe rain. At this moment several
had left the field as well as the General;
but the drums beat to arms, which caused
those who were absent to repair instantly to
the camp, and the lines were immediately
formed.

Captain Thornton's company was embodied with the matrosses, who were thought
too weak; and this was a great disappointment to him, whose intention was to be in
the front, whenever an engagement should
take place. Metcalf played before them to
the field; but the flag cannon sinking in a
bog, Captain Thornton exclaimed, " D—n
this accident; we shall see no sport to-day:"
and leaving his troop to assist the matrosses
in bringing up the cannon to their station,
he rode up opposite to the horse which were
going

going to engage. The regiments of Hamilton and Gardner were put in the front; and the Highlanders, after firing their pieces, threw them down, and difcharged their piftols in the horfes' faces, which caufed them to retreat, much confufed: and on the Duke of Perth exclaiming aloud, "Although the horfe have given way, yet the work is not accomplifhed," the enemy purfued with their broad fwords, cutting down the men as they fled; and the horfes did great mifchief, by breaking through their own foot, the men crying out at the fame time, "Dear brethren, we fhall all be maffacred this day!" On their paffing the artillery, the Captain of the matroffes feeing their perilous fituation, gave orders for all the train horfes to be cut from the cannon. General Hufke at this time came up with three regiments, and engaged the left wing of the Highlanders, ordering the rear and centre to keep firing, and the front to referve. The rebels, as was their cuftom, after the difcharge of their pieces, flung them away, and advanced with their

H 3.   broad

broad fwords clofe up to the firft line ; when
the front inftantly fired, and being fo near,
did more than double execution ; which
caufed them to retreat, leaving a great number dead upon the fpot.

The General obferving a vaft body of the
rebels on the right, drew up his men nearer
Falkirk, and gave orders to keep the town
until morning : however, on examining the
powder, they had the mortification to find
that the heavy rains had damaged it to fuch
a degree, that but few pieces could be fired ;
and the village being open on all fides, was
a circumftance fo favourable to the enemy,
as induced that General to form the refolution of quitting the town with all expedition,
and march to Linlithgow, where there was
more fhelter under the walls, in cafe of an
attack.—This meafure was fully juftified by
the event ; for the enemy purfued fo clofely,
that many were taken by furprife, as, in
confequence of the order to keep the town
all night, feveral had gone into the houfes to
put off their wet cloathes ; and thofe who
were

were apprifed of the retreat had no fooner left the place, than the rebels took poffeffion, and made a great many prifoners, amongft whom were twenty of Captain Thornton's men, with the Lieutenant and Enfign.

Mr. Crofts, the Lieutenant, having eighty guineas in his pocket, begged to make Lord George Murray his treafurer; which office his Lordfhip accepted, and had afterwards the *generofity* to return him six!

Captain Thornton, alfo, was in one of the houfes, for the purpofe before-mentioned, but had not time fufficient to effect his efcape; and being alarmed by the bagpipes at the door, he retreated up ftairs: in a few minutes feveral of the rebels rufhed up, in fearch of the fugitives; when one of them came to the very room door behind which he had taken refuge, and overlooking him, faid, "Here are none of the rafcals here." The woman of the houfe having feen the Captain go up ftairs, went to him foon after, and opening a clofet door, entreated him to enter, which he did.—She then brought

a

a dreffer, and placed difhes, &c. upon it, which prevented all appearance of a door in that place; and fortunately there was no bed in the room. About ten minutes after he had been fixed in his new quarters, a great number of people, confifting chiefly of High-land officers, amongft whom was Secretary Murray, took poffeffion of the apartment, which being large, they propofed making ufe of for bufinefs during their flay.

We will there leave Captain Thornton, and return to Metcalf, who with the matrof-fes was retiring from the field of battle.

Knowing that two of his mafter's horfes had been left at a widow's houfe a fhort diftance from the town, he made his way to the place, with intent to fecure them. This woman had in the morning expreffed great feeming loyalty to King George; but when Metcalf returned in the evening, the wind had changed :—She now extolled Prince Charles, and faid the defeat of *George's folk* was a juft judgment.

Metcalf

Metcalf went into the ſtable and found the horſes, ſaddled them, and was leading out the firſt, when he was ſurrounded by a few ſtragglers of the Highland army: " We muſt have that beaſt," ſaid they ; but Metcalf refuſing to give him up, they ſaid to one another, " Shoot him." On hearing two of them cock their pieces, he aſked, " What do you want with him ?"—They anſwered, that they wanted him for their Prince : " If ſo, you muſt have him," replied he. They took him, and immediately went off. Metcalf then brought out the other ; but as he was about to mount, the Captain's coach-man (whoſe name was Snowden) joined him, and Metcalf inquiring of him the fate of his maſter, was anſwered, that he had not ſeen him ſince he left the artillery, when he rode up with the horſe which were going to engage : this induced them to think that the worſt had befallen him. They then thought it adviſable to attempt falling in with the rear of the army, and endeavoured to ſlant the ground for that purpoſe ; but before they
had

had proceeded many yards, their horfe funk
up to the faddle-fkirts in a bog: however,
being ftrong and plunging out, they mounted
again, and foon joined it as they wifhed;
where on making diligent inquiry after their
Captain, they were told that he was left
behind; on which Snowden returned as far
as he could with fafety, but without gaining
any intelligence, and Metcalf walked on with
the army.

They arrived at Linlithgow, where they
halted; and the next day they marched to
Edinburgh. There the mob, and lower
orders of people, were very free in their
expreffions, and fome of the higher alfo
fpoke very warmly, in favour of Prince
Charles; making it appear clearly, *by their
own account*, that nothing could prevent his
coming to the Crown.

The next morning as many of Captain
Thornton's men as had efcaped being taken
prifoners, (about forty-eight in number,)
affembled; and none of them being quite
certain of having feen the Captain fince he
left.

left them with the cannon in the bog, they
fuppofed him to have fhared the fate of
many other brave men who had fallen in the
action of that day, and which they all fin-
cerely lamented—not only on account of the
favours he had individually conferred on
them, but for the great and liberal example
which he had invariably fhewn to his brother
officers, in the care and attention which he
paid to his men in general; the natural confequence of which was, that he poffeffed the
love and confidence of the foldiery. The
difappearance, alfo, of the two other officers,
and twenty of their men, greatly difpirited
them; and, together with the fufpenfion
from their regular pay, induced fome of
them to apply to Metcalf for a fupply, in
order to carry them home; but he laudably
refufed any aid he might have afforded
them, on being informed of the purpofe for
which it was required.

The army being fixed at Edinburgh, the
head-quarters were at the Abbey. The
fuperior officers fent for Metcalf, thinking it

a

a fingular circumftance that a perfon de-
prived of fight fhould enter into the army;
and knowing that his mafter was miffing,
they defired to converfe with him. One of
the officers belonging to the dragoons who
retreated from Falkirk fpeaking ironically of
Thornton's men, afked Jack how *he* got off
the field of battle.—Metcalf anfwered, " I
found it very eafy to follow by the found of
the dragoon horfes, they made fuch a *clatter*
over the ftones." This reply made the gentlemen turn the laugh againft him. Colonel
Cockayne likewife afked how he durft venture
into the fervice, blind as he was; to which
he replied, " that had he poffeffed a pair of
good eyes, he would never have come there
to have rifked the lofs of them by gunpowder." Then making his obeifance, he
withdrew: For Metcalf, though he had not
read books, had read *men ;* and received his
knowledge from the fchool of the world.

He now determined upon a journey to
Falkirk, in fearch of his Captain ; but this
being attended with difficulty, he applied to a
Knaref-

Knaresborough man who lived at Edinburgh
and was of the rebel party, telling him that
he wished to be a musician to Prince Charles,
as he found it was all over with the English.
The man informed him that they had a spy,
an Irishman, going to the Prince; on which
Metcalf set forward with him, and he pro-
mised to recommend him on their arrival at
Falkirk; but on coming up to the English
out-sentries, they were stopped :—Metcalf
inquired for the Captain, and informed him
of the real cause of his journey: by him he
was kindly advised to lay aside his project,
and told that he would lose his life; but still
persisting, he proceeded with the spy, and
arrived at Linlithgow, where they stayed all
night. They met with several women who
had been upon plunder, and were then on
their return to Edinburgh; and the spy
instructed them how to avoid the English
sentries. Metcalf was very careful to examine the cloathes they had got, thinking
that by chance he might meet with some of
his Captain's, ignorant as he was of his fate.

I   One

One of the women fent a token by Metcalf to her hufband, who was Lord George Murray's cook: this woman's guide was a horfe-dealer, who foon became acquainted with Metcalf, having frequented the fairs in Yorkfhire; and at this time by fome means had got introduced to the heads of both armies, and obtained a protection from each to prefs horfes occafionally.—This man's fate was remarkable; for going into Stirling, where the King's army lay, he found that orders were given to let no ftrangers pafs without an examination, which he underwent, and faid that he had a protection from General Hufke: being ordered to produce it, he had the misfortune to take that out of his pocket which he had got from the Pretender; and when informed of his miftake, inftantly produced the other—but too late; for he was tied up by the neck to a lamp-iron, without giving him time to put off his boots.

A fhort time before Metcalf and the fpy left the 'Change-houfe at Linlithgow, fome of the van guard of the rebels came in, and called

called for whiſkey ; and it was ſuppoſed that
they dropped there a ſilver-mounted piſtol,
which, on their ſetting out, the ſpy picked
up, and offered to Metcalf ; but he refuſed
it, ſaying, he thought it not proper to have
fire-arms about him, as he expected to be
ſearched : ſo they purſued their journey
and preſently fell in with the rebels out-
guard, ſeveral of whom accoſted Metcalf,
and as all ſeemed well, they were allowed
to paſs, and arrived at Falkirk, where he
inquired for Lord George Murray's cook,
to deliver his preſent, and was afterwards
introduced to and converſed with his Lord-
ſhip, Secretary Murray, and other gentlemen.   Lord George gave him part of a
glaſs of wine, an article at that time of great
value ; for as the rebels had been there three
times, and the Engliſh twice, they had almoſt
ſwept the cupboard clean of its crumbs.

Whilſt converſing with them, he was very
circumſpect, knowing that his life was in
danger, if the real purpoſe of his journey
ſhould be diſcovered.

He then made his way towards the market-place, where a number of Highlanders were assembled.—This was on Wednesday the 22d; but it happened that his master had left the place that morning, about four hours before his arrival.

We will now return to Captain Thornton, whom we left on Friday in the closet, in close neighbourhood with the Highland Chiefs, who every day transacted business in the room. The Quarter-Masters of the rebel army having taken the house, had given the woman to whom it belonged a small apartment backward; but every night she took care to carry him such provisions as she could convey through a crevice at the bottom of the door; and this mode she used for fear of alarming those who slept in the adjoining rooms. The closet was only a yard and a half square; and the Captain's cloathes being wet when he entered, made his situation the more uncomfortable, as he had got a severe cold, and sometimes could not forbear coughing, even when the rebels were in their room. Once in

in particular, hearing a cough, they faid one
to another " what is that ?" but one of them
anfwered, that it was fomebody in another
room ;—not in the leaft fufpecting a door in
the place where the clofet was.

On Monday night the woman of the houfe
went to the door to carry provifions as ufual,
when the Captain faid to her, " I am determined to come out, let the confequence be
what it may; for I will not die like a dog
in this hole ;" but fhe begged that he would
bear his confinement till the next night, and
fhe would adopt fome plan to effect his efcape.
She accordingly confulted an old carpenter,
who was true to the Royal caufe, and he came
the next night, removed the dreffer, and
liberated the Captain. They proceeded
down ftairs in the dark, to the woman's
apartment, where fhe made tea, whilft the
carpenter concerted their plan of operation.
They dreffed him in a pladdie and brogues,
with a black wig, and the carpenter packed
him up a bag of tools, as if he was going
with his mafter to work as foon as it was

I 3   light.

light. The Captain had only ten guineas about him, (having loſt his caſh with his Lieutenant, Mr. Crofts) eight of which he gave to the woman who had ſo faithfully preſerved him, and two to the carpenter, who, to ſecrete them, put them into his mouth along with his tobacco, fearful of a ſearch by the Highlanders, who would have ſuſpected him had they found more than a ſhilling. Every thing being ready, they ſet out, the Captain with his bag of tools following his ſuppoſed maſter. On coming into the croud, he looked about, and was rather behind ; and although in diſguiſe, did not look like a common workman ;—which making the old man dread a diſcovery, he called out to him, " Come alang, ye filthy loon : ye have had half a bannock and a mutchkin of drink in your wame—we ſhall be too late for our day's wark." Whether this artifice ſerved him or not, is uncertain ; but they got ſafe through the throng, and, leaving the high-road, purſued their journey acroſs the country. Having come to a riſing ground,

ground, the Captain took a view of Falkirk
moor, and faid, " Yonder's the place where
fuch a fad piece of work was made of it on
Friday laft." The old man at the fame time
looking the other way, faw two or three
hundred Highlanders, who had been on
plunder, coming down a lane which led from
Callendar-Houfe (Lord Kilmarnock's feat)
into the main road; and being defirous of
paffing the end of this lane before they came
up, in order to avoid them, faid, " We fhall
have a worfe piece of work of it than we had
on Friday, if you do not haften your pace;"
and begged the Captain to come forward,
which he did; but walking brifkly up a hill,
he fuddenly ftopped, and faid, "I am fick:"
however they gained their point, and paffed
the Highlanders; for had they come up with
them, the leaft injury would have been a
march back to Falkirk, as prifoners. On
going two miles farther, they arrived at a
houfe belonging to a friend of the carpenter's,
and which had been plundered : there the
old man got an egg, but not being able to
find

find a pan to boil it, he roasted it in peat-
ashes, and gave it to the Captain, to put in
his *wame*, for so he called his stomach.
Proceeding a few miles farther, they arrived
at another house, where they procured a
horse for the Captain.—He arrived at the
English out-posts, and making himself known
was permitted to pass, and reached Edinburgh in safety.

With respect to Metcalf, whom we left at
Falkirk, as his dress was a plaid waistcoat
laced with gold, which he had borrowed of
a friend at Edinburgh, together with a blue
regimental coat faced with buff, he told the
Highlanders, in answer to their inquiries,
that he had been fiddling for the English
officers, and that they had given him that
coat, which had belonged to a man who was
killed; and also that his intention was to
serve in the same capacity with Prince
Charles.—But a person coming up who had
seen Jack at Harrogate, said, "That fellow
ought to be taken up, for he has something
more than common in his proceedings;" on
which

which Metcalf was taken to the guard room,
and fearched for letters, but none were
found, having only a pack of cards in his
pocket, which they fplit, to fee whether they
contained any writing in the folds, but finding none, he was put into a loft in the roof
of the building, (where the fnow came in
very much) along with a dragoon, and fome
other prifoners, where for three days they
were fuffered to remain in confinement.

In a fhort time Metcalf and his fellow-
prifoners were tried by a court-martial, at
which he was acquitted, and had permiffion
given to go to the Prince; but wanting to
borrow a clean fhirt, they afked him where
his own were; he faid, at Linlithgow, but
that he durft not go there, on account of
George's devils. They told him that he
might fafely go with the Irifhman he came
with. He knew that his companion had
letters for the Highlanders' friends at Edinburgh, but had no intention to pafs the Englifh fentries. Metcalf amufed him with
affurances that he had ten pounds at Edinburgh,

burgh, for which he ſhould have no occaſion if he joined the Prince, and that he
might have the greateſt part of it: the ſpy,
on this, became extremely deſirous cf his
company to Edinburgh, wiſhing to finger the
money, and propoſed going acroſs the country; but Metcalf ſaid that *he* could paſs the
Engliſh ſentries, by ſaying that he was going
to Captain Thornton. They then proceeded,
and after going two miles, they met an cffi-
cer, who was reconnoitring, and he knowing
Metcalf, told him that his maſter was arrived
at Edinburgh, which news was highly plea-
ſing to him. On leaving the officer, the ſpy
accoſted him with "So, what you *are* going
to him."—"No," ſaid Metcalf, " nor to
any ſuch fellows." They then paſſed the
ſentry, as Metcalf propoſed, and arrived at
Edinburgh, where they parted, but promiſed
to meet the next evening at nine o'clock.
Jack went directly to his Captain, who re-
joiced at ſo unexpected a meeting. Metcalf
told him that he had given him a great deal
of trouble; adding, that he thought people
might

might come home from market without fetching —The Captain fmiled, and faid, " What is to be done, for I have neither money or cloathes, having left all behind at Falkirk ; but I have bills upon the road to the amount of three hundred pounds ?" This proved fortunate ; for had they been a few days fooner, they might by chance have been loft alfo ;—but the reafon of this delay was, that all letters, directed to Scotland, were at this time fent to London, to be examined at the General Poft Office. Metcalf told the Captain that he could get him fome money, which the other thought impoffible : how-ever he went to a known friend, and ob-tained thirty pounds.—Taylors were inftantly fet to work, and next morning the Captain was enabled to vifit his brother officers at the Abbey.

The army ftill quartered at Edinburgh, while part of the rebels were in Falkirk, and another part at Stirling, where they raifed feveral batteries, and befieged Stirling Caftle. The governor, General Blakeney, made little

oppo-

oppofition; and a fhot from the batteries
killing two or three men, fome of the officers
were greatly enraged, and threatened to
confine the Governor: But a little time
fhewed that he was right in his conduct;
for letting the rebels come pretty near the
walls, on a fudden he began fo hot a fire, as
to kill feveral of their men, demolifhing their
batteries, and difmounting their guns, which
made them glad to retreat, and raife the
fiege: and the General having deftroyed the
bridge, they were obliged to make a circuitous march before they were able to ford the
river.

The Duke of Cumberland arrived at Edinburgh on the 30th of January, 1746; and
two days afterwards marched out at the head
of the army, towards Falkirk, the rebels
leaving it a little time before. Captain
Thornton vifited the Duke often: his Royal
Highnefs took notice of Metcalf, and fpoke
to him feveral times on the march, obferving
how well by the found of the drum he was
able to keep his pace. On the army's arrival

val at Linlithgow, intelligence was received
that the rebels were on their march to give
them battle ; upon which the army was
drawn up in order, and the Duke rode
through the lines, and addreſſed the men as
follows : " If there be any who think them-
ſelves in a bad cauſe, or are afraid to engage,
thinking they may fight againſt any of their
relations, let them now turn out, receive
pardon, and go about their buſineſs, without
any farther queſtion."—On this, the whole
army gave three huzzas. But the intelligence proving falſe, they proceeded to Falkirk, and continued their route through Stirling, Perth, Montroſe, Briffin, and Stonehive,
to Aberdeen, where they halted. The rebel
army lay now at Strathbogie.

At Aberdeen the Duke gave a ball to the
ladies, and perſonally ſolicited Captain Thornton for his fidler, there being at that time no
muſic in the army except Colonel Howard's,
(the Old Buffs) and which being wind muſic
were unaccuſtomed to country dances. As
the rebel army was only twenty miles diſtant,

no

no invitations were fent till five o'clock, tho'
the ball was to begin at fix. Twenty-five
couple danced for eight hours, and his Royal
Highnefs made one of the fet, and feveral
times, as he paffed Metcalf, who ftood on a
chair to play, fhouted "Thornton, play up:"
but Jack needed no exhortation, for he was
very well practifed, and better inclined.

Next morning the Duke fent him two
guineas; but as he was not permitted to
take money, he informed his Captain, who
faid, that as it was the Duke's money, he
might take it; but obferved, that he fhould
give his Royal Highnefs's fervants a treat.
He had only three fervants with him, (viz.
his gentleman, cook, and groom.) So the
next night two of them paid Metcalf a vifit,
and a merry party they made, the Captain
ordering them great plenty of liquor.

In a little time they proceeded on their
march, and engaged the rebels on Culloden
moor, giving them a total defeat, although
they had targets to ward off the bayonet,
whilft they cut away with their broad fwords,
yet

yet the Duke found a method of fruſtrating
their plan, by puſhing the bayonet over the
right arm, which rendered their targets of
no uſe. Kingſton's Light Horſe purſued
them in their diſorder and flight, and made
a great ſlaughter amongſt them.

The Engliſh priſoners were now all liberated —Two or three of Captain Thornton's
men had died in priſon; and the reſt returned home.

The rebellion being completely ſuppreſſed,
Captain Thornton returned home alſo, accompanied by Metcalf, of whoſe family it is
full time to take ſome notice.—He had the
happineſs to find his faithful partner and
children in good health; and although the
former confeſſed that ſhe had entertained
many fears for her poor blind adventurer,
yet knowing that a ſpirit of enterprize made
a part of his nature, ſhe was often comforted
by the hope, that he would, in ſome degree,
ſignalize himſelf, notwithſtanding the misfortune under which he laboured.—This declaration, following a moſt cordial reception,

K 2   gave

gave full confirmation to an opinion which Metcalf had ever held, viz. that the careffes and approbation of the fofter fex, are the higheft reward a foldier can deferve or obtain.

The notice with which the Duke of Cumberland had honoured Metcalf, gave him much reafon to believe, that, had he followed him to London, he would have received more marks of his Royal favour.—But Metcalf was deficient to himfelf in this inftance; negleding to folicit further notice till it was judged too late to make application.

About a year after their return, a vacancy happening in the reprefentation for the city of York, the citizens fent for Mr. Thornton, and unanimoufly eleded him, free of all expence.

A fhort time after this, the militia was raifed, and he was, as his merit well entitled him to be, appointed Colonel of the Weft-York battalion; which fituation he held, with advantage to the fervice, and honour to himfelf, for the remainder of his life.

Blind

Blind Jack being now at liberty to choofe his occupation, attended Harrogate as ufual; but having, in the courfe of his Scotch expedition, become acquainted with the various articles manufactured in that country, and judging that fome of thofe might anfwer for him to traffic with in England, he repaired, in the fpring, to Scotland, and fupplied himfelf with various articles in the cotton and worfted way, particularly Aberdeen ftockings. For all thofe articles he found a ready fale at the houfes of gentlemen in the extenfive County of York; and being perfonally known to moft of the families, was ever very kindly received. He never was at a lofs to know, amongft a thoufand articles, what each had coft him, from a particular mode of marking.

It was alfo cuftomary with him to buy horfes, for fale in Scotland, bringing back galloways in return; and in this traffic he depended on feeling the animals, to direct his choice.

K 3 He

He alſo engaged pretty deeply in the con-
traband trade, the profits of which were at
that time much more conſiderable than the
riſk.

One time in particular, having received a
preſſing letter from Newcaſtle upon-Tyne,
requiring his ſpeedy attendance, he ſet out
on horſeback from Knareſborough at three
in the morning, and got into Newcaſtle in
the evening about ſix o'clock, the diſtance
nearly ſeventy-four miles, and did not feel
the leaſt fatigued.

Having received ſome packages, he employed a few ſoldiers to convey them to a
carrier, judging that men of their deſcription
were leaſt liable to ſuſpicion. After ſending
off his goods, he ſtayed two nights with ſome
relations he had there, and then ſet off for
home. He had with him about an hundred
weight of tea, caſed over with tow, and
tightly corded up; this he put into a wallet,
which he laid acroſs his ſaddle.

Coming to Cheſter-le-Street, (about half-
way between Newcaſtle and Durham) he
met

met at the inn an excifeman, who knew him as foon as he had difmounted, and afked him what he had got there. Metcalf anfwered, " It is fome tow and line for my aunt, who lives a few miles diftant ;—I wifh fhe was far enough for giving me the trouble to fetch it." The officer afking him to bring it in, he replied, " For the time I fhall ftay it may as well remain on the horfing-ftone." By this feeming indifference about his package, he removed fufpicion from the mind of the excifeman, who affifted in re-placing it acrofs the faddle ; when he purfued his journey, and got home in fafety.

Once having difpofed of a ftring of horfes, he bought, with the produce, a quantity of rum, brandy, and tea, to the amount of 200l. put them on board a veffel for Leith, and travelled over-land, on foot, to meet the veffel there. He had about thirty miles to walk, and carried near five ftone weight of goods which he did not choofe to put on fhipboard. At Leith he had the mortifica-tion to wait fix weeks, without receiving any
tidings

tidings of the veffel, which many fuppofed to
have been loft, there having been a ftorm
in the interval. The diftrefs of mind refult-
ing from this, induced him once to fay, "If
fhe is loft, I wifh I had been in her ; for fhe
had all my property on board." Soon after,
however, the fhip got into Leith harbour.
He there went on board, and fet fail for
Newcaftle ; but another ftorm arifing, the
mate was wafhed overboard, the mainfail
carried away, and the fhip driven near the
coaft of Norway. Defpair now became
general ; the profpect of going to the bottom
feeming almoft certain. He now reflected
on the impiety of his wifh refpecting the
former ftorm ; and fo effectually was his
way of thinking changed, that had he had all
the current coin of the univerfe, he would
have given it to have been on fhore. It now
appeared to him a dreadful thing to leave the
world in the midft of health and vigour ; but
the wind changing, hope began to return,
and the Captain put about for the Scotch
coaft, intending to make Arbrothie. A fignal

nal of diſtreſs was put up, but the ſea ran ſo
high, that no boat could venture out with a
pilot. He then ſtood in for the harbour,
but ſtruck againſt the pier end, owing to the
unmanageable ſtate of the veſſel, from the
loſs of her mainſail: ſhe narrowly eſcaped
being bulged; but having got to the back
of the pier, was towed round into the harbour, with near five feet water in her hold.
Her eſcape from the mercileſs elements,
however, did not ſeem to terminate her
dangers, the country people ſhewing a diſpoſition to ſeize her as a wreck, and plunder
her; but fortunately there was at hand a
party, conſiſting of an officer and twenty
men, of Pulteney's regiment, who had been
in purſuit of ſome ſmugglers; and Metcalf
knowing them well, (Colonel Thornton's
company being attached to that regiment)
the officer ſent three files of men to protect
the veſſel, while the crew were removing the
goods to a warehouſe.

As this veſſel ſtood in need of repairs,
Metcalf put his goods on board another,
and

and in her got to Newcaſtle. There he met
with an acquaintance; and from the ſeem-
ing cordiality at the meeting, he thought he
might have truſted his life in the hands of
this man. With this impreſſion, Metcalf
opened to him the ſtate of his affairs; informing him that he had got four hundred
gallons of gin and brandy, for which he had
a permit, and about thirty gallons for which
he had none, and which he wanted to land;
telling him, at the ſame time, of the harraſs-
ing voyage he had juſt finiſhed: But, it
ſeems, his misfortunes were only about to
commence; for, in a quarter of an hour, he
found that the man whom he had taken for
a friend had gone down to the quay ſide,
and, giving information of what he knew,
had all the goods ſeized, and brought on
ſhore. Metcalf imagined that none were
ſeizable but the ſmall part for which he had
not obtained a permit; but was ſoon undeceived, the whole being liable to ſeizure,
as not agreeing with the ſpecified quantity.

He

He then repaired to the Cuftom-Houfe,
and applied to Mr. Sunderland, the Collector.
This gentleman knew Metcalf, (being in the
habit of vifiting Harrogate) and received
him very kindly; but informed him, with
much concern, that it was not in his power
to ferve him, the captors being the excife
people, and not of his department —He,
however, fuggefted, that fome good might
refult from an application to Alderman
Peireth, with whom Metcalf was acquainted,
and who was particularly intimate with the
Collector of the Excife. The good Alderman gave him a letter to the Collector; re-
prefenting, as inftructed by Metcalf, that the
bearer had bought four hundred gallons of
fpirits, at the Cuftom-Houfe at Aberdeen;
and that the extra quantity was for the purpofe of treating the failors and other friends,
as well as for fea-ftock for himfelf. At firft
the Collector told him that nothing could be
done for him, until he fhould write up to the
Board, and receive an anfwer; but Metcalf
remonftrating on the inconvenience of the
delay,

delay, and the other re-confidering the letter, he agreed to come down to the quay at four o'clock in the afternoon, which he accordingly did, and releaſed every thing without expence.

A ſhort time after the regiment called the Queen's Bays were raiſed, they were quartered at Knareſborough and the adjacent towns; but, after a ſhort ſtay, they were ordered to the North. The country people ſeemed extremely unwilling to ſupply carriages for conveying the baggage; the King's allowance being but nine-pence a mile, per ton; that cf the County, one ſhilling in the Weſt Riding, and fifteen-pence in the North Riding. Metcalf having two waggons, (one of them covered) had a mind to try this new buſineſs; and, to make ſure of the job, got the ſoldiers to *preſs* his two carriages, which were accordingly loaded, himſelf attending them to Durham. Previous to loading, however, the country people, who knew the advantage of carrying for the army, and who had kept back, in hopes of an advance in the price,

price, came forward with their waggons, in oppofition to Metcalf; but the foldiers would employ no other.

Arriving at Durham, he met Bland's Dragoons, on their march from the North to York: they loaded his waggons again for Northallerton, and would willingly have engaged them to York; but this he was obliged to decline, having promifed to bring twenty-three wool packs to Knarefborough. He was juft fix days in performing this journey; and cleared, with eight horfes and the one he rode, no lefs a fum than twenty pounds; though many people were afraid to travel with foldiers.

Some time after the Queen's regiment had got to Durham, it received the ufual annual recruit of four horfes to a troop. The regiment having been fo lately raifed, had no old horfes: neverthelefs, four were to be fold from each. Metcalf had notice fent him of the fale, but did not receive the letter until the day before it commenced.— He fet off, however, that afternoon, for Durham, and riding all night, got there by day-break.   L   His

His firſt buſineſs was to become acquainted with the farriers; ſo getting about half-a-dozen of them together, and plying them heartily with gin, he began to queſtion them as to the horſes which were to be ſold.

Amongſt the number to be diſpoſed of, was a grey one, belonging to one of the drums. The man who had the charge of him not having been ſufficiently careful in trimming him, had burnt him ſeverely, which cauſed a prodigious ſwelling. Had this careleſs conduct been known to his ſuperiors, he would have been puniſhed for it: upon that account the matter was huſhed up. Metcalf, however, being apprized of the real cauſe, in the courſe of his converſation with the farriers, determined to purchaſe him, judging that they would be deſirous to part with him at any price; and in this conjecture he was not miſtaken.

The ſale began by bringing out ſeven bay horſes; ſix of which a gentleman bought for a carriage, and Metcalf purchaſed the ſeventh.

They

They then brought forward the grey horfe with his fwelled fheath, which excited many jokes and much laughter among the fpectators.—Our chapman bought him alfo, at the very low price of 3l. 15s. od. which was firft affixed by the auctioneer, but which, however, the people faid was very much beyond his value.

Having ufed fuch applications as he thought efficacious for his recovery, by the time he had got him home he had the fatiffaction to find him perfectly found; and within a week afterwards refufed fifteen guineas for him.—He kept him many years as a draught-horfe; and the other horfe alfo was fold to a profit, by which he thought himfelf very well paid for his journey to Durham.

In the year 1751 Metcalf commenced a new employ:—He fet up a ftage-waggon between York and Knarefborough, being the firft on that road, and conducted it conftantly himfelf, twice a week in the fummer feafon, and once in winter; and this bufinefs,

L 2   toge-

together with the occafional conveyance of
army baggage, employed his attention until
the period of his firft contracting for the
making of roads, which fuiting him better,
he difpofed of his draught, and intereft in
the road, to one Guifeley.

An act of Parliament having been obtained to make a turnpike-road from Harrogate to Boroughbridge, a perfon of the name
of Oftler, of Farnham, was appointed furveyor; and Metcalf falling into company
with him, agreed to make about three miles
of it, viz. between Minfkip and Fearnfby.—
The materials were to be procured from one
gravel-pit for the whole length: he therefore
provided deal boards, and erected a tempo-
rary houfe at the pit, took a dozen horfes to
the place, fixed racks and mangers; and
hired a houfe for his men at Minfkip, which
was diftant about three-quarters of a mile.
He often walked from Knarefborough in the
morning, with four or five ftone of meat on
his fhoulders, and joined his men by fix
o'clock: and by the means he ufed, he completed

pleted the work much fooner than was expected, to the entire fatisfaction of the furveyor and truftees.

During his leifure hours he ftudied mea-
furement in a way of his own; and when
certain of the girt and length of any piece of
timber, he was able to reduce its true contents to feet and inches; and would bring
the dimenfions of any building into yards or
feet.

Near the time of his finifhing this road,
the building of a bridge was advertifed
to be contracted for, at Boroughbridge;
and a number of gentlemen met for that
purpofe at the Crown inn there. Metcalf,
amongft others, went alfo. The mafons
varied confiderably in their eftimates. Oftler, the furveyor of the roads, was appointed
to furvey the bridge; and Metcalf told him
that he wifhed to undertake it, though he
had never done any thing of the kind before.
On this, the furveyor acquainted the gentlemen with what Metcalf had propofed; when
he was fent for, and afked what he knew

L 3   about

about a bridge: he told them, that he could
readily deſcribe it, if they would take the
trouble of writing down his plan, which was
as follows : " The ſpan of the arch, 18 feet,
being a ſemi-circle, makes 27 : the arch-
ſtones muſt be a foot deep, which if mul-
tiplied by 27, will be 486 ; and the baſes will
be 72 feet more.—This for the arch: it will
require good backing ; for which purpoſe
there are proper ſtones in the old Roman
wall at Aldborough, which may be brought,
if you pleaſe to give directions to that effect."
The gentlemen were ſurpriſed at his readi-
neſs, and agreed with him for building the
bridge. The perſons who had given in their
eſtimates, were much offended ; and as the
ſtone was to be procured from Renton, a
ſale-quarry belonging to one of the maſons
who were there, he was unwilling to ſell any
to Metcalf ; upon which he went to Farnham, and found good ſtones, which the
lime-burners had left, (being too ſtrong for
their purpoſe,) got them dreſſed at the place
for little money, conveyed them to Boroughbridge,

bridge, and having men to take them off the carriages, fet them, and completed the arch in one day; and finished the whole in a very short period.

Soon after, there was a mile and an half of turnpike-road to be made between Knaresborough-Bridge and Harrogate, which Metcalf alfo agreed for. Going one day over a place covered with grafs, he told his men that he thought it different from the ground adjoining, and would have them try for ftone or gravel, which they immediately did, and found an old caufeway, fuppofed to have been made in the time of the Romans, which afforded many materials proper for the purpofe of making the road. Between the Foreft-Lane head and Knarefborough-Bridge, there was a bog, in a low piece of ground, over which to have paffed was the nearest way; and the furveyor thought it impoffible to make a road over it: but Metcalf affured him that he could readily accomplish it.— The other then told him, that if fo, he fhould be paid for the fame length as if he had gone round.

round. Jack fet about it, caft the road up, and covered it with whin, and ling; and made it as good, or better, than any part he had undertaken. He received about four hundred, pounds for the road and a fmall bridge which he had built over a brook called Stanbeck.

There was an old houfe at Harrogate, with fome fields belonging to it, and denominated a meffuage, which was of more value, as having common-right upon the Foreft. It belonged to an old woman, and at her deceafe to her hufband. Metcalf went to the latter, and bought his contingent right in the houfe and land; and the old woman came to him foon after, to fell her life-eftate in it alfo. They agreed; and including both the net fum amounted to eighty pounds. In about three weeks after this purchafe, he fold it for upwards of two hundred pounds.

A road being projected between Harrogate and Harewood-Bridge, fix miles in length, a meeting was held, (the late Lord Harewood, then Mr. Lafcelles, being one of the

the party) to contract with any perfon who might be thought proper to make it. A great number of eftimates were delivered, but Metcalf obtained the contract. It was to be completed before the winter fet in ; and being a ftiff-clay foil, it was judged expedient to caft the whole length before they began to ftone it : on thefe accounts he agreed with the gentlemen, that no carriages fhould pafs whilft the road was making; and, by way of prevention, had fluices cut at each end of the lane, and wooden bridges, which he took up occafionally, thrown acrofs, for his own carriages to pafs over with the materials. He alfo hired two houfes, at a diftance from each other on the road, to entertain ftrangers who travelled on horfe-back, and the people employed in the under-taking, as there were not fufficient in the country. The fhort period he had contracted to complete the work in, obliging him to ufe the readieft methods, he had a wheel-plough drawn by nine horfes through the foreft, as the beft and moft expeditious way to get up

the

the roots of whin and ling, in parts where
they were ftrong; and being obliged to fuper-
intend the progrefs of the work, he obtained
leave from the innkeepers at Harrogate to
engage a fubftitute in his abfence. He completed his contract in the time allowed, to
the fatisfaction of the gentlemen truftees, and
of the furveyor; and received for the work
twelve hundred pounds.

There then being about a mile and an half
of road to be made through part of Chapel-
Town to Leeds, Lord Harewood and other
gentlemen met at the Bowling-Green in
Chapel-Town, to receive eftimates;—and
Metcalf got the contract. He alfo widened
the arch of Sheepfcar-Bridge; and received
for that and the road together near four
hundred pounds.

Between Skipton and Colne in Lancafhire
there were four miles of road to be made,
and eftimates were advertifed for. A number of gentlemen met, and Metcalf's pro-
pofals had the preference. The materials
were at a greater diftance, and more difficult

to

to be procured, than he expected; and a wet feafon coming on, made this a bad bargain; yet he completed it according to contract.

He next engaged for two miles on the Burnleigh road, which he completed; but was not more a gainer.

He then agreed for two miles of road which lay through Broughton to Martin; and two miles more which lay through Addingham, and over part of Romell's Moor. The fame truftees acted for thofe roads, as for that of Colne. Thefe he completed, and received one thoufand three hundred and fifty pounds from Mr. Ingham of Burnleigh and Mr. Alcock of Skipton.

After this, a meeting was held at Wakefield, to contract for making part of the road between that town and Halifax.—Metcalf engaged for four miles which lay between Mill Bridge and Belly-Bridge; and finifhed this alfo, though it was an extremely wet fummer.—He then took three miles more which lay between Belly-Bridge and Halifax, and completed it.—And alfo agreed

for

for five miles which lay between Wakefield and Checkingley-Beck, near Dewſbury.

The truſtees of the road, (Sir Rowland Winn, —— - Smith and W. Roebuck, Eſqrs.) meeting at Wakefield, to let part of the road leading to Pontefraƈt, and likewiſe from Wakefield to Doncaſter, Metcalf took three miles and an half which lay between Hag-Bridge and Pontefraƈt, and one mile and an half on the Doncaſter road, from Crofton through Foulby; all which he completed. The road from Wakefield, to Pontefraƈt, Doncaſter, and Halifax, being under the management of one company of truſtees, Mr. Allen Johnſon was treaſurer for one part, John Mills, Eſq; for another, both of Wakefield; and Mr. Valentine Stead, and Mr. William Cook, for the other parts: By the payment of theſe four gentlemen he received ſix thouſand four hundred pounds.

A road was then advertiſed to be let from Wakefield to Auſterland, intended to lead through Horbury, Almondbury, Hudderſfield, Marſden, and Saddleworth. A meeting

ing was held at Huddersfield, for the purpose. Sir John Kaye, Colonel Radcliffe, 'Squire Farrer, and several other gentlemen attended, and Metcalf agreed with them from Black-Moor Foot to Marsden, and from thence to Standish Foot; also from Lupset-Gate, thro' Horbury, about two miles and an half. At that time none of the road was marked out, except between Marsden and Standish Foot, leading over a common called Pule and Standish: the surveyor took it over deep marshes; but Metcalf not expecting it to have been carried that way, thought it a great hardship, and complained to the gentlemen, alledging it would be a much greater expence: they answered, that if he could make a complete road, he should not be a loser; and they were of opinion, that it would be necessary to dig the earth quite out of the marshes, until they came to a solid bottom.—Metcalf, on calculating that each marsh, upon an average, being three yards deep, and fourteen broad, would make two hundred and ninety-four solid yards of earth

M   in

in every rood, which, to have carried away,
would have been extremely tedious and ex-
penfive—and not only fo, but that the road
lying Eaft and Weft, would fill with fnow in
winter, (as it ufually falls in that direction,
when the wind is in the North)—argued the
point privately with the furveyor and feveral
of the gentlemen : but they all feemed im-
moveable in their former opinion. Metcalf
then appeared at the public meeting, and
the fubject was again brought forward ; but
knowing that it would be to little purpofe to
hold a conteft with them, he faid, "Gentlemen, as you feem to have a great deal of
bufinefs before you to-day, it appears quite
unneceffary to trefpafs upon your time on
this head :—I propofe to make the road
over the marfhes, after my own plan ; and if
it does not anfwer, I will be at the expence
of making it over again, after your's :" which
was confented to. And as he had engaged
to make nine miles of the road in ten months,
he began in fix different parts, with near
four hundred men employed. One of the
                                        places

places was Pule and Standiſh common, which
he caſt fourteen yards wide, and raiſed in a
circular form. Water in ſeveral places ran
acroſs the road, which he carried off by
drains; but found the greateſt difficulty in
conveying ſtones to the places for the purpoſe, on account of the ſoftneſs of the ground.
Numbers of clothiers uſually going that way
to Huddersfield market, were by no means
ſparing in their cenſure, and held much
diverſity of opinion relative to its completion.
But Metcalf got the piece levelled to the end,
and then ordered his men to bind heather,
or ling, in round bundles, and directed them
to lay it on the intended road, by placing the
bundles in ſquares of four, and laying another
upon each ſquare, preſſing them well down.
He then brought broad wheeled carts, and
began to lead ſtone and gravel for covering.
When the firſt load was brought and laid on,
and the horſes had gone off in ſafety, the
company huzza'd from ſurpriſe. They completed the whole of this length, which was
about half a mile; and it was ſo particularly

fine,

fine, that any perfon might have gone over it in winter, unfhod, without being wet. This piece of road needed no repairs for twelve years afterwards. Having finifhed the nine miles within the limited time, he took three miles from Standifh to Thurfton Clough, which he completed;—and afterwards fix miles and an half from Sir John Kaye's feat to Huddersfield; and from thence to Long-royd and Bridge toll-bar, about a mile and an half;—alfo four bridges, their fpans twenty-four feet each; together with fix bridges, the fpans of which were nine feet each. When all this work was finifhed, (the gentlemen having promifed that he fhould be no lofer) a meeting was called, and Metcalf attended: he affured them that the work he had completed extra to his firft bargain, in the marfhes and other places, deferved five hundred pounds: after fome debate, he was allowed three hundred pounds; though it was well worth the firft-named fum. He had made about twenty one miles in the whole, for which he received four thoufand five hundred pounds.   In

In the building of bridges, where the foundations were bad, he laid on a fufficient thicknefs of ling where it could be got, otherwife of wheat ftraw; he next laid planks five inches thick, with fquare mortifes cut through; and driving in a number of piles, he made the foundation fecure.—He then laid fprings for the arch upon the planks, which caufed all to fettle regularly when the weight came on. And though he built many arches, of different fizes, by taking this method none ever fell.

He undertook three turnpike-houfes upon the Wakefield and Aufterland roads, and completed them all. Believing there was a ftone quarry on the South Eaft fide of Huddersfield, in ground belonging to Sir John Ramfden, he bored fecretly in the night-time before he undertook the road, and was fuccefsful in finding it. After finifhing the road, having fome vacant time, and having likewife difcovered the quarry, Sir John gave him liberty to lead away ftone. He now took houfes to build, particularly one belong-

M 3   ing

ing to Mr. Marmaduke Hebdin, nine yards wide, twenty-three yards long, and twenty-one feet from the foundation to the fquare of the building;—it had twenty chimnies or pipes : And this he alfo completed.

He undertook the road from Dock-Lane head, in Yorkfhire, to Afhton-under-Line, in Lancafhire ; alfo from the guide poft near the latter place, to Stockport, in Chefhire ; and alfo between Stockport and Mottram-Longdale : the whole length being eighteen miles. He fet men to work in different parts, with horfes and carts to each company ; and though he loft twenty horfes in one winter, he was not difcouraged ; obferving that " horfe leather had been dear a long time, but he hoped now to reduce the price." Notwithftanding this misfortune, he completed the whole, including a great number of drains and arches, which were all done to the fatisfaction of the truftees and furveyor ; and received for the work four thoufand five hundred pounds.

He

He then took eight miles more which required feveral drains and arches.—He raifed
one hollow nine yards, and built fufficiently
on each fide to keep up the earth, with battlements on the top; for which he received
two thoufand pounds.

One day being met by Sir Geo. Warren,
who inquired if he was at leifure, and being
anfwered in the affirmative, he defired to fee
him at his houfe at Poynton. Metcalf went,
and agreed to make about five miles of a
private road through the Park. He took
twelve or fourteen horfes of his own, and
brought large quantities of ftone into Sir
George's grounds, for the ufe of draining.
For this he received feveral hundred pounds,
and great favours alfo from Sir George and
his lady.

A road was to be made between Whaley
and Buxton, in Derbyfhire, to avoid a hill:
it went over a tedious piece of ground called
Peeling Mofs; the whole road being four
miles in length, with fome part ftrong rock,
which was to be blafted with gunpowder.—

In

In taking this road, Metcalf met with ſtrong
oppoſition from a ſon of one of the commiſ-
ſioners; but Peter Legh, Eſq; of Lyme, and
another gentleman, ſupporting Metcalf, he
gained the point, and completed the under-
taking, with ſeveral drains and fence walls;
for which he received near eleven hundred
pounds.

He next took a mile and an half of High-
Flats, between Huddersfield and Sheffield;
and finiſhed it likewiſe, to the amount of
three hundred pounds.

Eight miles of road were next advertiſed
to be made between Huddersfield and Halifax. A meeting was held, and ſeveral perſons attended with eſtimates for making it.
One part was very rocky, and full of hollows,
and the ground in a very bad ſituation, particularly between Elland and Salterſhebble,
and through a place called Grimſcar Wood,
which was very boggy and rough. Many
were of opinion that it was impoſſible to
make a road over that ground. But by
building up the hollows, and lowering the
hills,

hills, Metcalf accomplished it:—And it is worthy of remark, that he never undertook any road which he did not complete, altho' he has taken many which persons who had their sight durst not engage in. He finished the road, with a great number of fence walls and drains, to the satisfaction of the surveyors and trustees, and received for it two thousand seven hundred and eleven pounds.

A little after this, a road was advertised to be made between Congleton and the Red-Bull Inn, in Cheshire, about six miles in length; but the materials were about three miles distant in several places. A meeting for letting this road was held at a place called Auderſley, which Metcalf attended; and being a stranger in that part, he fortunately met with three gentlemen who knew him, viz. —— Clows of Macclesfield, —— Downs of Sigleigh, and —— Wright of Mottram, Esqrs. two of them Justices of the Peace.— They said to the trustees, "Gentlemen, you have only to agree with this man, and you may be assured of having your work well done."

The

The road, however, was no let that day, the bufinefs being deferred until another meeting to be held at Congleton, where Metcalf and others attended with eftimates.——
"Gentlemen," faid Metcalf, "I am a ftranger to you, and you may with reafon queftion my performing the bargain; but to prevent any doubt, I will firft do one hundred pounds worth of work, and afterwards be reafonably paid as it goes forward; the hundred pounds may lay in the treafurer's hands till the whole is completed, and then to be paid." On this propofal, and the three gentlemen's recommendation at the former meeting, they agreed with him, although there was an eftimate given in lower than his by two hundred pounds. He completed the road, to the great fatisfaction of the furveyor and truftees, and received three thoufand pounds.

During the time that Metcalf was engaged in making this road, having one day occafion to ftop at Congleton, he met, at the Swan inn there, one Warburton, a capital farmer, who lived about a mile diftant. This man
was

was remarkable for fporting large fums in various ways, and no fooner faw Metcalf, than he accofted him thus : "I underftand that you play at cards."—Metcalf replied, "Sometimes, but not often;" being much furprifed that a ftranger fhould know he had that propenfity. Warburton offered to play him for five or ten pounds, the beft of five games at put ; but this he thought fit to decline : in the prefence of his friends he would not have feared to play for twenty ; but being in a ftrange place, and having a large undertaking relative to the turnpikes, he concluded that it would be highly im- prudent to game. The farmer, however, perfifting in his defire for play, Metcalf, after a little confideration, determined to try the effect of ridicule on his new acquaint- ance, faying, " I have not now time ; but if you will meet me here this day fortnight, I will play you, the beft of five games, for a leg of mutton, four pennyworth of cabbage, and five fhillings worth of punch." The farmer, pleafed with any profpect of engaging

him,

him, agreed to the wager, and infifted that
the money fhould be depofited with the landlord; which was accordingly done. During
the interval, Warburton fpread the ftory of
his engagement to play with a blind man;
and, thinking it a good joke, invited many
of his friends to the entertainment. Metcalf
came at the time fixed, having firft engaged
a friend from Buxton to accompany him,
whofe chief bufinefs it was to fee that his
adverfary did not play tricks with the cards.
Three guineas to two were offered to be laid
on Warburton; and Metcalf's friend obferving this, expreffed a wifh to take the
odds, if agreeable to him: to this, Metcalf
replied, that he meant only to amufe himfelf
by playing for mutton and cabbage; and,
that if any fums were laid, he would forfeit
his wager. When all parties were affembled, Metcalf, out of joke, propofed to his
adverfary to club for all the articles, and
treat the company; but this he pofitively
refufed, alledging that he had collected his
friends for the purpofe of feeing the match
played.

played. On this, Metcalf called to the landlord for a fiddle, and playing on it for a
little while, was afked by the farmer what he
meant: " To enable you," faid he, " to tell
your children, that when you played with a
blind man, you *played to fome tune!*" They
then went into a large room, and were followed by a crowd of people, amongft whom
were two Juftices of the Peace, and feveral
clergymen. The game now began, and
Metcalf won the two firft; his adverfary got
the third, and pulling out his purfe, offered
to lay five guineas on the rubber: this was a
tickling offer to Metcalf, but having refolved
againft playing for money, he made fhift to
overcome the temptation. Metcalf won the
next game; and, of courfe, the rubber. On
this the farmer laid a large fum on the table,
and offered to play for the amount; but
Metcalf would only play for liquor, for the
good of the company. The farmer agreeing,
they began again, and Metcalf prefently won
two games, when a gentleman prefent fhewed
a great defire to play with him for money,

N but

but in vain ; fo winning this rubber alfo, he
faddled his antagonift with the whole fcore,
and not fatisfied with the triumph already
gained, began to banter him forely on his
childifh manner of playing, and telling him,
that when the road work fhould ceafe for the
Chriftmas holidays, he would come to his
houfe, and teach him to play like a *man*.

The quantity of liquor yet to come in
being large, detained many of the company
until five in the morning ; and Warburton,
who had got pretty drunk, by way of com-
fort, declared before parting, that of twenty-
two fine cows, he would rather have loft the
beft, than have been beaten fo publicly.

Metcalf apprehending that he might now
be folicited by many to engage in play, and
confidering the importance of his other en-
gagements, called afide Mr. Rolle, the furveyor of the road, and begged of him to give
fixpence, upon condition of receiving five
pounds, if he (Metcalf) fhould play any
more at cards for eighteen months, the time
allotted to finifh the road. Mr. Rolle appro-
ving

ving highly of this, they returned to the company, and Metcalf making the propofal, received the furveyor's fixpence publicly; and thus put an end to all further importunity.

Here Metcalf finds it his duty to fufpend, for a while, his road-making narrative, to introduce, for the laft time, the mention of the much loved Partner of his cares, whom he had brought into Chefhire, and left at Stockport, that fhe might avail herfelf of the medical advice of a perfon there, famed for the cure of rheumatic complaints, of which defcription her's was thought to be :—But human aid proving ineffectual, fhe there died, in the fummer 1778, after thirty nine years of conjugal felicity, which was never interrupted but by her illnefs or his occafional abfence.

In his treatment of her, Metcalf never loft fight of the original diftinction in their circumftances, always indulging her to the utmoft that his own would allow; but fhe had no unreafonable defires to gratify. She died in the fixty-firft year of her age, leaving four children; and was buried in Stockport church-yard. N 2 In.

In 1781 the road between Wetherby and
Knaresborough was let.——He undertook that
part which led through Ribston and Kirk-
Deighton, till it joined the great North road
leading from Boroughbridge to Wetherby;
and also built two toll-houses upon the road;
and received about three hundred and eighty
pounds.

Metcalf had a daughter married in Cheshire,
to a person in the stocking business. The
manufacturers in this line, in the neighbourhood of Stockport, talked of getting loads of
money; and Metcalf thought that he would
have a portion of it also: he accordingly got
six jennies and a carding engine made, with
other utensils proper for the business; bought
a quantity of cotton, and spun yarn for sale,
as several others did in the country. But it
cost him much trouble and expence, before
he got all his utensils fixed: the speculation
likewise failed; and a time came when no
yarn could be sold without loss. Then Metcalf got looms, and other implements proper
for weaving calicoes, jeans, and velverets:——
for having made the cotton business an object
of

of particular attention, he was become very well acquainted with the various branches of it. He got a quantity of calicoes whitened and printed, his velverets cut, dyed, &c. and having fpun up all his cotton, he fet off with about eight hundred yards of finifhed goods, intending to fell them in Yorkfhire, which he did at Knarefborough and in the neighbourhood ; and his fon-in-law was to employ his jennies until he came back. On his return, coming to Marfden near Huddersfield, where he had made a road fome years before, he found that there was to be a meeting, to let the making of a mile and an half of road, and the building of a bridge over the river that runs by the town, fo as to leave the former road, in order to avoid the fteepnefs of a hill. At the perfuafion of fome of his friends, he ftaid till the meeting, and agreed with the truftees. The bridge was to be twelve yards in the fpan, and nine yards in breadth. Thefe too he completed, and received a thoufand pounds ; but the feafon being wet, and the ground over which he had to bring his materials very fwampy, and at a diftance from the road, he loft confiderably by it. In

In 1789 he was informed that there was a
great quantity of road to be let in Lancashire:
he accordingly went, and took a part between
Bury and Eslington, and another part from
Eslington to Ackrington; as also a branch
from that to Blackburn. There were such
hollows to fill, and hills to be taken down, to
form the level, as was never done before: in
several of the hollows the walls were ten yards
high, before the battlements were put on the
top. He had two summers allowed to finish
this work in; but the trade in Lancashire
being brisk, made wages very high, and the
navigation at that time cutting through the
country so employed the men, that it was a
very difficult matter to procure a sufficiency
of hands. The first summer the rains were
so perpetual, that he lost about two hundred
pounds; but in the next he completed the
whole work, and received by the hands of
Mr. Carr of Blackburn three thousand five
hundred pounds; and, after all, was forty
pounds loser by it.

In the year 1792 he returned into Yorkshire; and having no engagement to employ
his

his attention, he bought hay to fell again, meafuring the ftacks with his arms; and having learnt the height, he could readily tell what number of fquare yards were contained, from five to one hundred pounds value. Sometimes he bought a little wood ftanding; and if he could get the girth and height, would calculate the folid contents.

From that period he has fettled on a fmall holding at Spofforth, near Wetherby; and his houfe is kept by a daughter and fon-in-law.

At Chriftmas, 1794, he paid a vifit to the prefent Colonel Thornton, and his mother, at Thornville-Royal; and the reception he met with was fuch as fully reminded him of former days at Old Thornville, where he had fpent many Chriftmaffes. The truly refpectable Relict, and the worthy Reprefentative, of his late Commander, always receive Blind Jack with a condefcending affability, highly flattering to one in his humble ftation.

Having known the ftreets of York very accurately in the earlier part of his life, he determined, on the commencement of the laft year, to vifit once more that ancient city,

where

where he had not been for the fpace of thirty-
two years : He found alterations for the better
in Spurriergate, Blakeftreet, the Pavement,
&c. and being now in the neighbourhood of
Middlethorp, where he had, in the year 1735,
fpent a half-year fo happily, he refolved to
have another *look* at it, in the poffeffion of its
prefent worthy mafter. From Mr. Barlow's
houfe there is a road which leads to Bifhop-
thorpe ; and this road he clearly recollected,
though fixty years had elapfed fince he had
gone that way before : fo retentive was his
memory on this occafion, that he difcovered
an alteration in the hanging of two gates by
a wall-fide near the above manfion. At Mr.
Barlow's he ftaid feveral nights, which, he
fcarcely need add, were fpent moft agreeably,
he endeavouring to make his fiddle fpeak the
fatisfaction and hilarity felt by its owner.
Returning to York, he fpent a few nights at
the houfe of another friend ; and fetting out
on the 10th of January, 1795, he walked to
Green-Hammerton, in his way to Thornville-
Royal, in about three hours and an half, being

ten

ten miles; proceeded to Thornville that night, and to Knaresborough next morning the 10th, which being the birth-day of Sir Thomas Slingsby's eldest son, and which was kept with the utmost festivity, he resolved to spend at the worthy Baronet's. Here he closed the festive season of Christmas, after a tour of some weeks amongst his friends;— to whom, in particular, he submits, with the utmost deference, this imperfect Sketch of a Life, with which only can terminate his grateful remembrance of their numerous favours.

F I N I S.